Model-based Geostatistics for Global Public Health

Methods and Applications

CHAPMAN & HALL/CRC
Interdisciplinar y Statistics Series

Series editors: N. Keiding, B.J.T. Morgan, C.K. Wikle, P. van der Heijden

Recently Published Titles

For more information about this series, please visit: https://www.crcpress.com/go/ids

Model-based Geostatistics
for Global Public Health

Methods and Applications

Peter J. Diggle
Emanuele Giorgi

CRC Press
Taylor & Francis Group
Boca Raton London New York

CRC Press is an imprint of the
Taylor & Francis Group, an **informa** business

A CHAPMAN & HALL BOOK

CRC Press
Taylor & Francis Group
6000 Broken Sound Parkway NW, Suite 300
Boca Raton, FL 33487-2742

First issued in paperback 2021

Version Date: 20190125

ISBN-13: 978-1-03-209364-2 (pbk)
ISBN-13: 978-1-138-73235-3 (hbk)

To Mandy and Iulia

Contents

Preface

This book provides an introductory account of model-based geostatistics and its application in public health research.

The term *geostatistics* is a short-hand for the collection of statistical methods relevant to the analysis of geolocated data, in which the aim is to study geographical variation throughout a region of interest but the available data are limited to observations from a finite number of sampled locations. This scenario is typical of applications in low-resource settings where comprehensive disease registries do not exist. Accordingly, most of the examples in the book relate to public health research in low-to-middle-income countries, drawing on our experience of collaborative work in Africa, Asia and South America.

Geostatistical methods originated in the South African mining industry in the early 1950s (Krige, 1951). They were subsequently developed, by Georges Matheron and colleagues at Fontainebleau, France, into a self-contained methodology for addressing problems of spatial prediction; see Matheron (1963) or, for a book-length account, Chilès & Delfiner (2016). This methodology has subsequently been applied in many different fields, spanning the social, physical and health sciences. Watson (1971, 1972) first pointed out the connection between geostatistics and classical stochastic process prediction. The books by Ripley (1981) and Cressie (1991) subsequently placed geostatistics within the more general setting of statistical methods for spatially referenced data. Diggle et al. (1998) coined the term *model-based geostatistics* to mean the application of general principles of statistical modelling and inference to the analysis of geostatistical data. In particular, they emphasised the use of likelihood-based inference within an explicitly declared parametric model, typically a generalized linear mixed model (Breslow & Clayton, 1993) with a latent spatial process included in the linear predictor.

The R software environment (`www.r-project.org`) has become the standard vehicle for disseminating new statistical methodology as open-source software through the provision of R *packages* as add-ons to the basic R language. Packages are made available through the CRAN repository, which is accessible via the R project web-page, `https://cran.r-project.org/`. All of the analyses reported in this book can be reproduced using the R package `PrevMap` and its predecessor `geoR`. R scripts are provided on the book's web-site, `https://sites.google.com/site/mbgglobalhealth/`.

Many of the public health applications described in the book fall under the general heading of *disease mapping* problems. A basic scenario is the following. How can we best use data on empirical prevalences of a disease of interest at a

set of sampled locations within a designated region A to construct a map of the
spatial variation in prevalence throughout A? Many variations on this basic
scenario arise according to the practical focus of particular applications. What
is the relationship between disease risk and exposure to one or more spatially
varying risk-factors? Where do unexplained "hot-spots" occur within A? How
is the spatial distribution of prevalence changing over time? What would be
an efficient spatial sampling design for monitoring changes in prevalence over
time?

The remainder of our applications concern *exposure mapping*, i.e. constructing spatially continuous maps of potential exposure to risk-factors, such
air pollutant concentrations, from a spatially discrete network of measurement
sites. Similar questions are relevant in this context, and can again be answered
using geostatistical methods.

Our aim has been to write a book that is accessible not only to statisticians
but also to students and researchers in the public health sciences. Those in the
latter category may initially struggle with some of the mathematical formalism
that we use in describing the various statistical models and methods. However,
we believe that the effort involved in becoming comfortable with mathematical
notation, and with some basic concepts in probability and statistical inference,
is well worthwhile for at least three reasons. Firstly, expressing a statistical
model in mathematical terms forces precision of thought and explicit dec-
laration of underlying assumptions, both of which can be masked by vague
statements of the kind, "we fitted a regression model of disease risk on age,
gender and socio-economic status." Secondly, understanding the differences
amongst statistical testing, estimation and prediction helps to ensure that the
analysis of a set of data focuses on the correct scientific question. Finally, by
embedding geostatistical methods within a general inferential paradigm we
greatly reduce reliance on *ad hoc* methods and thereby ensure that our analyses are statistically efficient, i.e. within the declared model our inferences are
as precise as they can be.

To help this second category of reader negotiate any initial technical difficulties, we have included a brief account of the underlying statistical theory
and methods in Appendix A. Also, at the end of Chapter 1 we signpost those
parts of the book that less mathematically inclined readers may wish to skip
on a first reading. We emphasise that the reader needs only to understand the
statements of the various results, not how they are derived.

Conversely, statisticians may be less familiar than public health scientists
with software tools such as geographical information systems (GIS) for drawing the high-quality maps that are an essential part of communicating the
results of a geostatistical analysis to users. We have therefore included Appendix B, which describes how to use R packages to do this. We could have used
an open-source GIS instead. For example, we sometimes use the Quantum GIS
(QGIS) system (https://www.qgis.org/en/site/) in our own work. But we
think it is more helpful to the reader that we keep all of our analysis tools
within a single software environment.

In writing this book we have benefited greatly from discussion and collaboration with many friends, colleagues and students without whom this book would never have been written. Special thanks are due to Madeleine Thomson (Columbia University), who introduced PJD to the world of global public health research having spotted the potential for geospatial statistical methods to contribute to this important area of work.

The opportunity to make a contribution, however small, to public heath in some of the world's poorest countries has been both a humbling and an enormously rewarding experience for us both.

Peter J Diggle and Emanuele Giorgi, Lancaster, 17 November, 2018

List of Figures

List of Tables

1

Introduction

CONTENTS

1.1   Motivating example: mapping river-blindness in Africa

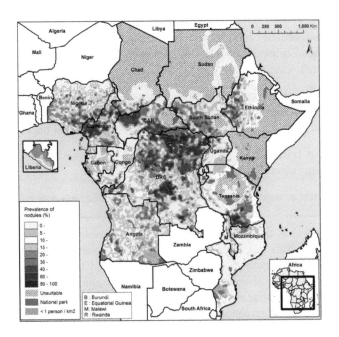

FIGURE 1.1
Map of estimated pre-control prevalence of onchocericasis infection Africa-wide.

1

The African Programme for Onchocerciasis Control (APOC) was a WHO-coordinated programme to control onchocerciasis, more commonly known as river-blindness; see `www.who.int/blindness/partnerships/APOC/en/`. The programme was launched in 1995 and covered 20 participating African countries (Coffeng et al., 2013). In 2015, APOC was absorbed into the more wide-ranging Expanded Special Project for the Elimination of Neglected Tropical Diseases (ESPEN), see `www.afro.who.int/en/espen.html`.

River-blindness is a parasitic infection transmitted by the bite of an infected blackfly. APOC's principal control strategy is annual mass prophylactic treatment of affected communities with an antifilarial medication, ivermectin, that kills the parasites before they can cause clinical disease. A useful tool to help prioritise mass distribution of the drug to the worst affected areas would be a map showing the geographical variation in prevalence throughout the APOC target region. Figure 1.1 shows such a map.

River-blindness

- **The disease.** River-blindness, also known as Onchocerciasis, is an infectious disease caused by the parasitic worm *Onchocerca volvulus*. It is currently endemic in 30 African countries, Yemen, and isolated regions of South America.

- **The vector.** It is transmitted from human to human through repeated bites of blackflies of the genus *Simulium* which breed along fast-flowing rivers.

- **The symptoms.** The subcutaneous dying larvae can cause long-term damage to the skin. The larvae residing in the eye can also lead to visual impairment and, in severe cases, to blindness.

- **The treatment.** The standard medication used to kill the larvae of an infected person is ivermectin.

- **Source.** `www.cdc.gov/parasites/onchocerciasis`

How was this map constructed? Field-epidemiologists visited a number of rural communities in each of the 20 countries, selected a sample of between 30 and 50 adult members of each community, and tested each sampled individual for presence/absence of infection using a rapid diagnostic test (REMO, `http://www.who.int/apoc/cdti/remo/en/`). The resulting data can be characterised as a set of triplets, $(m_i, y_i, x_i) : i = 1, ..., n$ in which m_i is the number of individuals tested and y_i the number who tested positive for infection in each of n sampled communities at locations x_i.

At any sampled location x_i, the observed proportion, $p_i = y_i/m_i$, of sampled individuals who test positive is an estimate of the local prevalence, $P(x_i)$. To extend these estimates to unsampled locations x we need to interpolate spatially between the estimates at the sampled locations. There are many ways in which this can be done. One of the simplest objective methods is inverse-distance-weighting (Shepard, 1968). To interpolate the prevalence, $p(x)$, at a location x, Shepard's method first calculates the distances d_i between each x_i and x, then defines a set of weights, $w_i(x) = 1/d_i^p$ where, typically, $p = 1$ or 2, and defines the interpolated prevalence as

$$p(x) = \left(\sum_{i=1}^{n} w_i p_i \right) \Big/ \left(\sum_{i=1}^{n} w_i \right).$$

Note that, by definition, any interpolation method will reproduce the observed prevalences p_i at the corresponding sampled locations x_i. From a statistical perspective, this is not obviously a good idea. We might prefer also to smooth the p_i themselves, to acknowledge that part of the variation amongst the p_i reflects binomial sampling error rather than genuine geographical variation in prevalence.

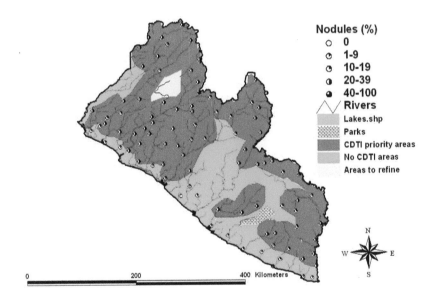

FIGURE 1.2
Prevalence map for onchocerciasis in Liberia produced by WHO (methodology unspecified).

In this book, our guiding principle is that problems of this kind are best solved by formulating, and empirically validating, a statistical model for the data and applying general principles of statistical inference to provide an answer to the scientific question, with the following features: it is as precise as possible; and it carries with it a quantitative indication of exactly how precise or imprecise the answer is. The details of how we do this will be developed in the remainder of the book. Here, we simply illustrate the result of applying our methods to data from one of the APOC countries, Liberia, and compare the result with a map published by the WHO that used an unspecified method of interpolation (http://www.who.int/apoc/cdti/remo/en/). The two maps use the same data, and are shown in Figures 1.3 and 1.2. On the model-based map, the data are plotted at their sampling locations with circles whose radius is proportional to the corresponding prevalence, and colour-coded to identify quintiles of the observed prevalences from blue (lowest) through green, yellow and brown to red (highest). The model-based estimates of prevalence are mapped using a continuous colour gradation from approximately 0.05 (5%, white) to 0.3 (30%, dark green).

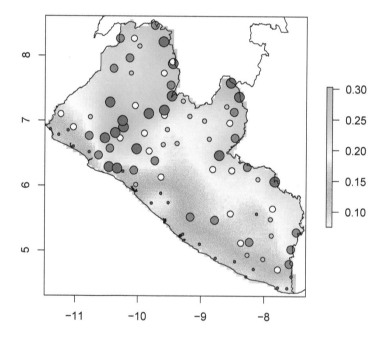

FIGURE 1.3

Prevalence map for onchocerciasis in Liberia produced by the application of model-based geostatistics.

On the WHO map, the data are shown as circles at sampled locations with filled black wedges indicating the corresponding observed prevalence. Areas on the map are coloured red where "onchocerciasis is highly endemic and constitutes a significant public health problem" and green where "the prevalence of skin nodules is ≤20%." Areas coloured yellow indicate "where results are not clear and additional rapid epidemiological assessment surveys are needed."

The two maps tell a qualitatively similar story. However, an advantage of the model-based map is that it comes with estimates of its precision. Note in particular that the WHO defines areas with prevalence greater than 20% to be "treatment priority areas." Accordingly, Figure 1.4 shows model-based predictive probabilities that local prevalence is greater than 20%. The maps in Figures 1.2 and 1.4 identify similar priority areas (red on the WHO map, green on the model-based map indicating a high probability that prevalence is greater than 20%) and a band of low-prevalence areas close to the coast. To a rough approximation, the yellow band in the model-based map, corresponding to a predictive probability 0.5, i.e. a fifty-fifty call, matches the boundary between the red and green zones on the WHO map. Also, the yellow area in the north of the country identifed by the WHO as needing refinement corresponds roughly to the yellow-green patch in the model-based map, indicating predictive exceedance probabilities in the range 0.5 to 0.6 . Another material difference between the two maps is that the model-based map is more nuanced. In particular, it indicates a gradation of risk moving away from the coast in the southern half of the country that contrasts with the WHO map's abrupt boundary between green and red areas. Finally, the model-based map finds little evidence to support the WHO's delineation of an isolated red zone in the south-east of the country.

1.2 Empirical or mechanistic models

The word "model" is widely used in science, but with subtly (or not-so subtly) different meanings. One useful distinction, albeit an over-simplification, is between *empirical* and *mechanistic* models. In any scientific modelling exercise, information can be acquired in two very different ways: from *data* and from *contextual knowledge*. Empirical and mechanistic models can be regarded as opposite ends of a continuum according to the extent to which data-based (empirical) and contextually-based (mechanistic) information is used in formulating the model.

Geostatistical modelling is typically empirical in character, but well-founded mechanistic assumptions can add value in any particular application nad, conversely, empirical elements can improve the fit of an over-simplified mechanistic model.

For example, consider the case of how we might model age-specific prevalence of an infectious disease antibody. Suppose that all people in a given community become sero-positive at a constant rate λ, whilst those who are

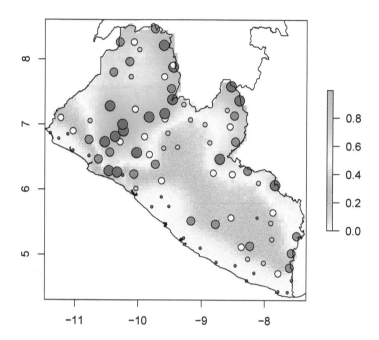

FIGURE 1.4
Predictive probability map for exceedance of 20% onchocerciasis prelevance
in Liberia, the WHO-defined threshold for "treatment priority areas."

already sero-positive revert to sero-negative at a rate ρ. Denoting the antibody prevalence at age t with $p(t)$, the mechanism of sero-conversion is then
driven by the following ordinary differential equation

$$\frac{dp}{dt} = \lambda(1 - p(t)) - \rho p(t),$$

whose solution is

$$p(t) = \frac{\lambda}{\lambda + \rho}[1 - \exp\{-(\lambda + \rho)t\}].$$

This mechanistic model is called the *reversible catalytic prevalence* model and
is often used to model malaria sero-prevalence data (Corran et al., 2007).

The assumption of a constant sero-reversion rate ρ is perhaps reasonable
as it relates to an internal biological process, but the assumption of a constant sero-conversion rate is more dubious because a person's susceptibility to

infection can be expected vary according to a variety of measured and unmeasured characteristics of their lifestyle and environment. This suggests adding an empirical sub-model for λ, for example a log-linear regression model,

$$\lambda_i = \exp(\alpha + \beta d_i + U_i),$$

where i denotes the individual, d_i is a measured characteristic and U_i is a random effect that we include in the model as a proxy for unmeasured characteristics of individual i.

1.3   What is in this book?

This book describes model-based geostatistics by guiding the reader through each of the steps of a geostatistical analysis that are illustrated with applications to a range of public health problems drawn from our own research.

Most of our appplications are set in low-to-middle-income countries where, as in our introductory example of onchocerciasis prevalence mapping in Liberia, geostatistical methods are particularly relevant because their key feature is their ability to predict an outcome of interest at locations where outcome data have not been obtained.

In high-income countries, health outcome data are typically recorded either as individual events linked to each patient's place of residence, or as counts of the numbers of events in each of a number of areal units that partition the region of interest.

Figure 1.5 is an example of the first kind. Each point on the map shows the residential location of a case of non-specific gastro-intestinal illness, resident in the English county of Hampshire, reported to a phone-based 24/7 triage service, NHS Direct, that operated UK-wide between 1998 and 2014. The data are an example of a spatial (strictly, spatio-temporal since the reporting date of each case is also available) *point process*. The density of points on the map largely reflects the underlying population distribution, and the question of primary interest is whether, and if so where and when, statistically unusual local concentrations of cases occur. Books that describe statistical methods for analysing spatial point process data include Diggle (1983), Ilian et al. (2008), Diggle (2013) and Baddeley et al. (2016) .

Figure 1.6, taken from López-Abente et al. (2006), is an example of the second kind. It shows the geographical variation in the relative risk of lung cancer mortality in the Castile-La Mancha Region of Spain. Here, the recording units are administrative areas that together partition the whole of Spain. The data are counts of the number of recorded lung cancer deaths in each area, and the mapped quantities are estimates of the relative risk of lung cancer as the cause of death. Data of this kind are variously called lattice (Cressie, 1991), discrete spatial variation (Gelfand et al., 2010) or spatial interaction (Besag,

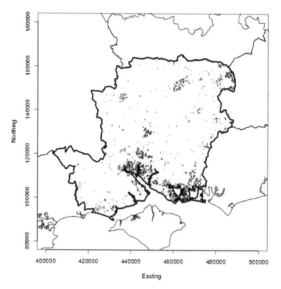

FIGURE 1.5
Residential locations (unit post-codes) of reported cases of nonspecific gas-
trointestinal illness in Hampshire, UK.

1974) data. Their distinguishing feature is that modelling and inference are
linked explicitly to the specified set of small areas.

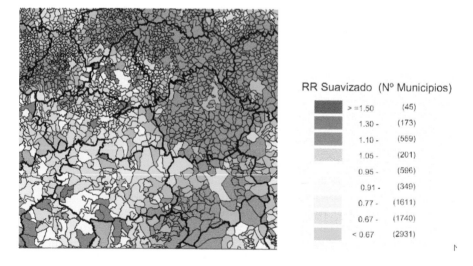

FIGURE 1.6
Estimated relative risk of lung cancer mortality in Castile-La Mancha, Spain.

Diggle et al. (2013) have argued that point process or lattice data can sometimes be analysed within the framework of geostatistics. However, geostatistical, point process and lattice data each have their own associated methodologies. For most of this book, we will limit our scope to statistical models and methods for geostatistical data as conventionally defined, i.e. data gathered at a discrete set of points in an area of interest, A, with the aim of understanding the behaviour of an unobserved, spatially continuous phenomenon that exists throughout A and could, in principle if not in practice, be observed at any point x in A.

Chapters 2 to 6 contain core material on linear and generalised linear geostatistical modelling and on geostatistical design. Chapters 7 to 10 describe various extensions to the core, and require the reader to have a stronger grasp of mathematical notation. In these later chapters, less mathematically inclined readers may wish to read Appendix A, which is a short review of the foundations of statistical inference, before tackling these chapters. They may also wish to skip the technical details on a first reading, and focus instead on the applications so as to gain an understanding of the underlying concepts.

For all the examples illustrated in this book, we have used the `PrevMap` package (Giorgi & Diggle, 2017) available for download from the Comprehensive R Archive Network (`cran.r-project.org/web/packages/PrevMap/`). Most of the data-sets and `R` code to enable the reader to reproduce, or modify, our analyses are available from the book's web-page, `sites.google.com/site/mbgglobalhealth`. Data-sets that are currently not available for download from the website will be made available as soon as permission is granted by the data-owners. In Appendix B we describe how to handle spatial data in `R` in order to produce high-quality maps. However, the substantive content of the book can be understood without any knowledge of `R`.

2

Regression modelling for spatially referenced data

CONTENTS

The starting point of almost any geostatistical analysis is an ordinary least squares regression analysis. This exploratory step of the analysis can provide useful insights on the empirical association between the available explanatory variables and the outcome of interest. Analysis of the residual from an ordinary least squares regression can also help to inform the formulation of an appropriate geostatistical model. In particular, residual analysis can and should question the assumption of independently distributed observations. In this chapter we describe the class of generalised linear models (GLMs), with a focus on Gaussian, Binomial and Poisson models. We then show how to carry out exploratory spatial analysis to look for evidence of residual spatial correlation. We also introduce the class of generalised linear mixed models to account for over-dispersion of nominally Binomial or Poisson data. Finally, we show how to use diagnostic procedures based on the empirical variogram to test for spatial independence. Text-book accounts of the theory and applications of linear and generalized linear models include Weisberg (2013) and Dobson & Barnett (2008), respectively.

2.1 Linear regression models

Consider a continuous response variable, Y_i, measured at a discrete set of locations, $\{x_i : i = 1, \ldots, n\}$, where each x_i lies within a geographical region of interest, A. A standard linear regression model for the data is then based on the following assumptions.

- *Independence.* The Y_i are mutually independent random variables.

- *Linearity.* $E[Y_i] = d(x_i)^\top \beta$, where $d(x_i)$ is a vector of explanatory variables with associated regression coefficients β.

- *Homoscedasticity.* $\text{Var}[Y_i] = \sigma^2$, for all $i = 1, \ldots, n$.

- *Error-free explanatory variables.* The explanatory variables $d(x_i)$ are measured without error and their observed values can therefore be treated as known constants.

- *Normality.* The Y_i follow a Gaussian distribution with mean and variance as defined above.

We can write this model in a more compact form as

$$Y_i = d(x_i)^\top \beta + U_i, \tag{2.1}$$

where U_i are independently and identically distributed Gaussian variables with mean zero and variance σ^2.

Johann Carl Friedrich Gauss

"Gaussian" is a synonym for "Normal" after the German mathematician Johann Carl Friedrich Gauss (1777-1855), who is generally credited with giving the first description of the Normal distribution and the method of least squares estimation in the course of solving a problem in astronomy.

FIGURE 2.1
Gauss and the Gaussian distribution depicted on a 10 Deutschmark banknote.

A geostatistical model relaxes the assumption of stochastic independence between the observations by allowing spatially correlated residuals. However, before questioning the independence assumption it is important to use the available covariates $d(x_i)$ to explain some of the variation in the measurements

Y_i. This has two potential benefits: exploiting any association between $d(x_i)$ and Y_i leads to more precise geostatistical predictions, both at and between data-locations; and regression adjustments often result in a simpler residual correlation structure.

The World Health Organization's child growth standards

The World Health Organization Multicentre Growth Reference Study (MGRS) was conducted between 1997 and 2003 with the objective of generating standard levels of child growth using anthropometric measurements taken from approximately 8500 children in Brazil, Ghana, India, Norway, Oman and the USA. Figure 2.2 shows the resulting standard growth curves for boys with height-for-age Z-score 0, ± 1, ± 2 and ± 3. The small bump at two years is due to the average difference between recumbent length and standing height, which are used to measure chidren below and above two years of age, respectively.
Source. www.who.int/childgrowth

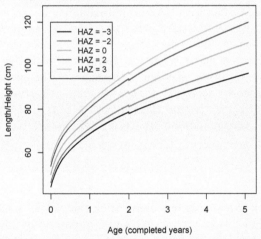

FIGURE 2.2
Curves of child-growth based on length-for-age (0-2 years) and height-for-age (2-5 years). Each curve correspond to a specific length/height-for-age Z-score value (HAZ) of -3, -2, 0, 2 and 3, with 0 being the standard level of growth.

2.1.1 Malnutrition in Ghana

We now use the linear regression model for an exploratory analysis of data on malnutrition in Ghana. The data are available on the book's web-page as the file malnutrition.csv. They contain information on the height-for-age

Z-score (HAZ) of 2,671 children. HAZ is an age-standardized measure of the deviation from standard child growth. Values of HAZ close to zero indicate normal growth, whilst values less than -2 are interpreted as an indication of stunted growth. In the data a group of households, called a *cluster*, is associated with a single GPS location. The objective of our analysis is to estimate the stunting risk in Ghana by mapping the predictive probability that HAZ is below -2. To do this, we develop a non-spatial linear regression model for HAZ in order to explore the empirical association of HAZ with the available explanatory variables.

Figure 2.3(a) shows a point-map of the cluster locations. To explain the variation in HAZ, we then consider the following covariates: the age of the child in years (d_1); the level of education of the child's mother (d_2), which takes integer values from 1 (poorly educated) to 3 (highly educated); the wealth index of the household (d_3), which takes value from 1 (poor) to 3 (rich).

Figure 2.3(b) shows the relationship between HAZ and age. We observe a negative association between HAZ and age in the first two years of life, which then becomes slightly positive beyond two years. Figures 2.3(c) and 2.3(d) show box-plots of HAZ for each of the categories of d_2 and d_3. In both cases, the observed relationship with HAZ is approximately linear, with higher levels of maternal education and wealth index associated with larger values of HAZ.

To capture the non-linear effect of age, we use a *broken stick*, which consists of a set of connected linear segments. Here, we break the stick at ages 1 and 2 years. The resulting algebraic expression for the age effect is $a(d_1) = \sum_{h=1}^{3} \beta_h b_h(d_1)$, where $b_1(d_1) = d_1$, $b_2(d_1) = max(0, d_1 - 1)$ and $b_3(d_1) = max(0, d_1 - 2)$. If $Y_j(x_i)$ denotes the HAZ of the j-th child at cluster location x_i, our non-spatial regression model for the data is now

$$Y_j(x_i) = \beta_0 + \sum_{h=1}^{3} \beta_h b_h(d_{1,ij}) + \beta_4 d_{2,ij} + \beta_5 d_{3,ij} + U_{ij}. \qquad (2.2)$$

The broken stick is a simple but useful trick that allows non-linear relationships between an explanatory variable and a response variable to be captured within a model that can be fitted using standard *linear* regression software. It is an example of a wider class of models known as *splines*; for a general discussuon, see for example De Boor (2001).

Table 2.1 reports the ordinary least squares estimates of the regression coefficients in (2.2). The estimated effect of age on HAZ, as expressed by the *broken sticks* regression coefficients, β_1 to β_3, is shown in Figure 2.4. We observe a sharp decline in HAZ in the first two years of life and a gradual recovery afterwards. The positive estimates of β_4 and β_5 are in accordance with the marginal association observed in Figure 2.3.

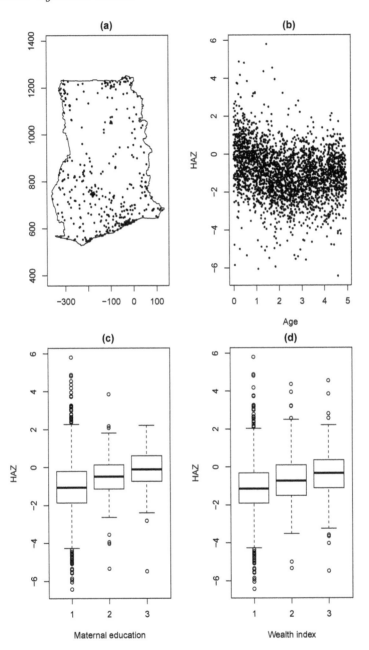

FIGURE 2.3
Plots for the data on malnutrition in Ghana: (a) point-map showing the geo-referenced locations of the sampled clusters; (b) scatter plot of height-for-age Z-scores against age; (c) box-plots of height-for-age Z-scores for each category of maternal eduction; (d) box-plots of height-for-age Z-scores for each category of household wealth score. The unit of distance between geographical locations in (a) is 1 km.

TABLE 2.1
Ordinary least squares estimates with nominal standard errors (Std. Error)
and 95% confidence intervals (CI) for the regression coefficients of the linear
regression model in (2.2).

Parameter	Estimate	Std. Error	95% CI
β_0	-0.644	0.112	(-0.864, -0.424)
β_1	-0.638	0.132	(-0.896, -0.379)
β_2	-0.205	0.206	(-0.608, 0.198)
β_3	0.963	0.115	(0.737, 1.188)
β_4	0.166	0.060	(0.048, 0.285)
β_5	0.361	0.037	(0.288, 0.434)

2.2   Generalised linear models

The class of generalised linear models (GLMs), introduced by Nelder & Wedderburn (1972), extends the linear regression modelling framework of the previous section to responses that are not Normally distributed. Of particular relevance in public health applications are the Binomial and Poisson regression models that are widely used to analyse counts of disease cases.

In a GLM, we continue to assume that observations are independent, and that explanatory variables are measured without error, but we allow the response Y_i to belong to any member of the family of exponential distributions; this family includes the Gaussian, Binomial and Poisson distributions as special cases. Also, we introduce a *link function*, $g(\cdot)$ such that

$$E[Y_i] = m_i g^{-1}(\eta_i)$$

where the known quantity m_i is called the *off-set* and $\eta_i = d(x_i)^\top \beta$ is the *linear predictor*. Finally, we assume that the variance of the response, $\mathrm{Var}[Y_i]$, is equal to, or at least is proportional to, a known function of the expectation, $E[Y_i]$. The Gaussian linear regression model is the special case in which $g(\cdot)$ is the identity function and $\mathrm{Var}[Y_i]$ is a constant, whilst the Poisson loglinear regression model corresponds to the link function $g(u) = \log(u)$ and $\mathrm{Var}[Y_i] = E[Y_i]$.

2.2.1   Logistic Binomial regression: river-blindness in Liberia

We now consider the data on skin nodule (SN) prevalence in Liberia, previously described in Section 1.1. Our objective is to identify "hotspots" as areas where SN prevalence is above 20%.

Our response variable, Y_i, is the number of people with SN out of n_i, the number sampled. A natural model for Y_i is a Binomial distribution with

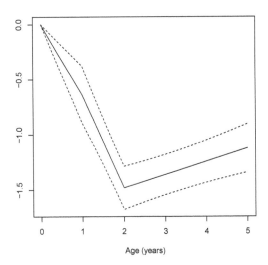

FIGURE 2.4
The estimated effect (solid line) of age on HAZ from model (2.2). This is
formally expressed by the *broken stick* in (2.2). The dashed lines are the 95%
confidence intervals.

number of trials n_i and probability of having SN $p(x_i)$. Because $p(x)$ is a
probability, a linear model for the relationship between $E[Y_i]$ and explanatory
variables is not a natural choice. The *logistic* model specifies that the log-odds
of $p(x)$ is linearly related to explanatory variables, corresponding to a linear
predictor $\eta_i = \log[p(x_i)/\{1 - p(x_i)\}]$, which ensures that $p(x)$ lies between 0
and 1, as it must. This is by no means the only available specification, although
it is the one most widely used in practice.

Figure 1.3 showed a trend in prevalence, increasing with distance from the
coast roughly along a north-east direction. To capture the dominant feature
of this trend we define the linear predictor as

$$\eta_i = \log\left\{ \frac{p(x_i)}{1 - p(x_i)} \right\} = \beta_0 + \beta_1 x_{i,1} + \beta_2 x_{i,2} \qquad (2.3)$$

where $x_{i,1}$ and $x_{1,2}$ are the coordinates of x_i on the east-west and north-south
axis, respectively.

In Table (2.2), the maximum likelihood estimates of the regression coefficients indicate that, on average, as we move 100 km eastward the odds become
about $\exp(\hat{\beta}_1 \times 10^2) \approx 13\%$ times larger, whilst moving 100 km northward
these become about $\exp(\hat{\beta}_2 \times 10^2) \approx 36\%$ larger.

TABLE 2.2

Maximum likelihood estimates with associated standard errors (Std. Error) and 95% confidence intervals (CI) for the regression coefficients of the Binomial regression model in (2.3).

Parameter	Estimate	Std. Error	95% CI
β_0	-4.245	0.449	(-5.124, -3.365)
$\beta_1 \times 10^3$	1.215	0.427	(0.378, 2.051)
$\beta_2 \times 10^3$	3.107	0.425	(2.274, 3.939)

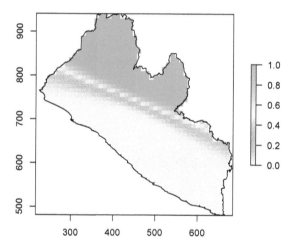

FIGURE 2.5

Predictive probability map for exceedance of 20% skin nodule prevalence in Liberia from the Binomial regression model in (2.3).

We can also use (2.3) to map the probability of exceeding the WHO-declared benchmark of 20% prevalence at any location x in Liberia. This exceedance probability is approximately

$$P\left([1 + \exp\{-(\hat{\beta}_0 + \hat{\beta}_1 x_1 + \hat{\beta}_2 x_2)\}]^{-1} > 0.2\right)$$
$$= \quad P(\hat{\beta}_0 + \hat{\beta}_1 x_1 + \hat{\beta}_2 x_2 > -\log 4), \tag{2.4}$$

which can be computed using the Gaussian approximation to the joint distribution of $(\hat{\beta}_0, \hat{\beta}_1, \hat{\beta}_2)$; we refer the reader to the appendices for more details on the properties of the maximum likelihood estimators $\hat{\beta}$.

The approximation in (2.4) stems from the substitution of estimates for the unknown regression parameters. In later chapters we discuss this in more detail and show how the approximation can be avoided by using predictive inference.

Figure 2.5 shows the resulting exceedance probability map for a 20% threshold, computed over a 10 by 10 km regular grid covering the whole of Liberia. The map shows high exceedance probabilities in the northern part of Liberia but does not capture any of the small scale variation shown earlier in Figure 1.4. Although the assumed linear assocation of longitude and latitude with the log-odds of prevalence does not appear completely to capture the geographical variation in prevalence, we will describe in later chapters how it can still be a useful component of a more flexible class of models.

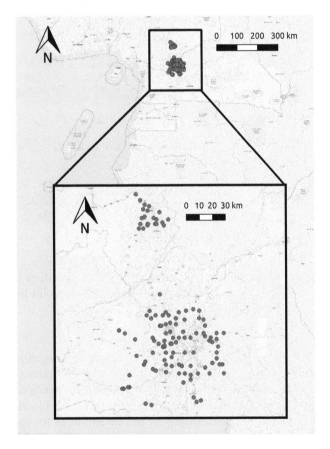

FIGURE 2.6
Map of the locations of the *Anopheles gambiae* counts in Southern Cameroon.

2.2.2 Log-linear Poisson regression: abundance of *Anopheles Gambiae* mosquitoes in Southern Cameroon

The data that we analyse in this section were extracted from a database developed by Tene Fossog et al. (2015), and are available from the book's web-page in the file **anopheles.csv**. They include counts of *Anopheles gambiae* mosquitoes at 116 geo-referenced locations in Southern Cameroon. The objective of the study is to map the spatial distribution of *A. gambiae* within the study area.

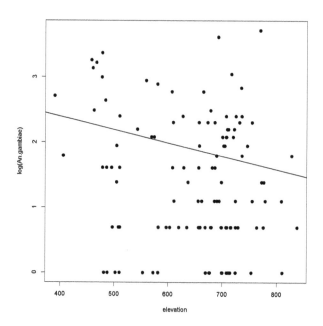

FIGURE 2.7
Scatter plot of the log-counts of mosquitoes against elevation for the data on *Anopheles gambiae* counts in Southern Cameroon; the solid line is the fitted mean using the Poisson log-linear model (2.5).

Figure 2.6 shows the point-map of the counts of mosquitoes. This shows a higher concentration of *A. gambiae* in the northern part of the sampled area than in the south. The survival of *A. gambiae* depends on temperature and humidity, as well as on their success in obtaining a blood meal. We use elevation as a proxy for the suitablity of the environment for successful breeding. Figure 2.7 shows the empirical relationship between the log-counts and elevation. This suggests that as elevation increases, the density of mosquitoes tends to decrease, although the association is rather weak.

Anopheles mosquito

- **General information.** *Anopheles* is a genus of mosquito comprising 460 species, of which 100 can transmit malaria. Among these, *Anopheles gambiae* is one of the most known species due its ability to transmit the most dangerous malaria parasite species, *Plasmodium falciparum*.

- **Geographical distribution.** Worldwide, except in Antarctica.

- **Habitat.** The larvae are mainly found in clean, unpolluted water including fresh- or salt-water marshes, mangrove swamps, rice fields, grassy ditches, edges of streams and rivers, and temporary rain pools. However, larval habitat varies, even within the same species, in relation to the amount of vegetation.

- **Source.** www.cdc.gov/malaria/about/biology/mosquitoes/

TABLE 2.3
Maximum likelihood estimates with associated standard errors (Std. Error) and 95% confidence intervals (CI) for the regression coefficients of the Poisson regression model in 2.5.

Parameter	Estimate	Std. Error	95% CI
β_0	3.210	0.214	(2.790, 3.630)
$\beta_1 \times 10^3$	-2.037	0.338	(-2.701, -1.374)

Let Y_i and $d(x_i)$ denote, respectively, the count of *A. gambiae* and the elevation, in meters, at location x_i. In a standard GLM analysis, we model Y_i as a set of mutually independent Poisson variables, with means $\lambda(x_i)$, such that

$$\log\{\lambda(x_i)\} = \beta_0 + \beta_1 d(x_i), \text{ for } i = 1, \ldots, 116. \qquad (2.5)$$

The maximum likelihood estimates are reported in Table 2.3. We estimate that for an increase of 1,000 meters in elevation the mosquito density reduces by about $(100 \times \{1 - \exp(-2.037)\} \approx)$ 87%. However, environmental factors other than elevation also affect the distribution of *A. gambiae*. This invites the question, how should we account for the effect of these unmeasured variables on the distribution of the Y_i?

2.3 Questioning the assumption of independence

In the two examples above, we have shown how to explore and model the empirical relationship between the outcome of interest and the available explanatory variables using generalized linear models. We now consider whether, and if so in what way, the unexplained variation in Y might invalidate the assumption of independently distributed observations. More specifically, we consider how to test for the presence of residual *spatial correlation*, an informal definition of which is that "close things are more related than distant things."[1]

To this end, we first extend the generalised linear modelling framework by assuming that Y_i conditionally on a set of independent Gaussian variables Z_i, with mean zero and variance τ^2, belongs to the family of exponential distributions, with link function $g(\cdot)$ and linear predictor

$$\eta_i = d(x_i)^\top \beta + Z_i. \tag{2.6}$$

More generally, the model extension is to the class of *generalized linear mixed models* (GLMMs), where the term "mixed" refers to the linear predictor containing both fixed effects, the regression component $d(x_i)^\top \beta$, and random effects, the Z_i.

Note that, in the special case of the linear Gaussian model, this is not an extension but only a re-expression of the model. This model is usually written in the compact form (2.1),

$$Y_i = d(x_i)^\top \beta + U_i, \tag{2.7}$$

where the U_i are mutually independent and Normally distributed, $U_i \sim N(0, \sigma^2)$. The equivalent definition as a generalized linear model is: $E[Y_i] = \eta_i$; $\eta_i = d(x_i)^\top \beta$; $Y_i \sim N(\eta_i, \sigma^2)$. The apparent extension to a generalized linear mixed model by setting $\eta_i = d(x_i)^\top \beta + Z_i$, where the Z_i are mutually independent and Normally distributed, as previously specified, and $U_i \sim N(0, \tau^2)$, leads to write

$$\begin{aligned} Y_i &= d(x_i)^\top \beta + U_i + Z_i \\ &= d(x_i)^\top \beta + Z_i^*, \end{aligned} \tag{2.8}$$

where now the Z_i^* are mutually independent and Normally distributed, $Z_i^* \sim N(0, \sigma^2 + \tau^2)$. It follows that the models (2.7) and (2.8) are empirically indistinguishable unless the value of either ν^2 or τ^2 is known from information external to the data, $Y_1, ..., Y_n$.

Let $[\cdot]$ be a shorthand notation for "the distribution of $\cdot$"; using the law of total variance (see text box), it follows that

$$\text{Var}[Y_i] = E[\text{Var}[Y_i|Z_i]] + \text{Var}[E[Y_i|Z_i]]. \tag{2.9}$$

In Section 2.2 we described the standard Poisson and Binomial regression

---

[1]This very simple idea is sometime called the First Law of Geography and is the foundation of the concept of spatial dependence. It was originally proposed by the geographer Waldo Tobler at a meeting of the International Geographical Union's Commission on Qualitative Methods, in 1969.

models, for which the term Z_i in (2.6) is absent. For the Poisson model, the expectation and variance of Y_i are equal, whilst for the binomial model the relationship between the expectation and variance is that $\text{Var}[Y_i] = E[Y_i](1-E[Y_i]/m_i)$ where m_i is the binomial denominator. When Z_i is present, the effect of the second term on the right hand side of (2.9) is to inflate the variance rerlative to the expectation, so that in the Poisson case $\text{Var}[Y_i] > E[Y_i]$, and in the Binomial case $\text{Var}[Y_i] > E[Y_i](1 - E[Y_i]/m_i)$. GLMMs can therefore be used to model *over-dispersion*, meaning that the data show greater variability than would be expected under the standard formulation as described in Section 2.2. Common causes of over-dispersion are the omission of risk factors, resulting in unexplained variation in the response variable, or intrinsic random variation that results from individual characteristics, such as behavioural or genetic traits. Note that under (2.6) the Y_i are still independently distributed, and the model therefore cannot account for omitted risk-factors that are themselves spatially structured. Nevertheless, (2.6) represents a useful starting point for investigating over-dispersion.

The laws of total expectation and variance

Let Y_1 and Y_2 be two random variables such that the variance of Y_1 is finite. The *law of total expectation* states that

$$E[Y_1] = E[E[Y_1|Y_2]].$$

Now if we consider the variance of Y_1 this can be written as

$$\text{Var}[Y_1] = E[Y_1^2] - (E[Y_1])^2$$

which, using the law of total expectation can be rewritten as

$$
\begin{aligned}
E[Y_1^2] - (E[Y_1])^2 &= E[E[Y_1^2|Y_2]] - (E[E[Y_1|Y_2]])^2 \\
&= E[\text{Var}[Y_1|Y_2] + E[Y_1^2|Y_2]] - (E[E[Y_1|Y_2]])^2 \\
&= E[\text{Var}[Y_1]] + [E[E[Y_1^2|Y_2]] - (E[E[Y_1|Y_2]])^2].
\end{aligned}
$$

From the last equation above, we obtain the so called *law of total variance* given by

$$\text{Var}[Y_1] = \text{Var}[E[Y_1|Y_2]] + E[\text{Var}[Y_1|Y_2]].$$

In the geostatistical setting, unexplained variation in Y_i might also include a spatially structured component, which would manifest itself in the form of residual spatial correlation. This leads to the question: how can we look for evidence of spatial dependence in the data? A natural starting point is to question the independence of the Z_i in (2.6). However, since the Z_i are not observable we need a way of making an informed guess, called a *point predictor*, of each of them. The point predictor can be obtained as a suitable summary of the the conditional distribution of Z_i given the data, called the *predictive distribution* of Z_i. Under the assumed model (2.6), this distribution is obtained by an application of Bayes' theorem, to give

$$[Z_i|y_i] = [Z_i, y_i]/\int_{\mathbb{R}} [Z_i][y_i|Z_i]\, dZ_i. \qquad (2.10)$$

For a general GLMM, (2.10) is analytically intractable, but can be evaluated using numerical methods.

Common choices for the point predictor are the mean, median or mode of $[Z_i|y_i]$. In the remainder of the book, we shall denote any of these point predictors of Z_i as $\hat{Z}_i$. A theoretical argument in favour of the mean is that this choise minimises the mean square error, $\mathrm{E}[(\hat{Z}_i - Z_i)^2]$. The distribution $[Z_i|y_i]$ depends on the model parameters. For *plug-in prediction* we replace them by their maximum likelihood estimates. We give a more detailed discussion of this in later chapters.

2.3.1 Testing for residual spatial correlation: the empirical variogram

We now define the *empirical variogram*, a tool for exploratory spatial analysis that we use here to test for residual spatial correlation after fitting a GLMM. The idea behind the variogram is that in the absence of spatial correlation, the the squared differences between pairs of predicted residuals, $(\hat{Z}_h - \hat{Z}_k)^2$, should fluctuate around a constant value, equal to twice the variance of the $\hat{Z}_h$, because $\hat{Z}_h$ and $\hat{Z}_k$ are independent, irrespective of the distance u between their corresponding locations. In the presence of residual spatial correlation in the data, we would expect the squared differences to be smaller on average at shorter distances u, as a result of the stronger correlation between $\hat{Z}_h$ and $\hat{Z}_k$ (Tobler's First Law of Geography).

We therefore define the *empirical variogram* of the predicted residuals $\hat{Z}_i$ as

$$\hat{V}(u) = \frac{1}{2|N(u)|} \sum_{(h,k)\in N(u)} (\hat{Z}_h - \hat{Z}_k)^2, \qquad (2.11)$$

where $N(u) = \{(i,j) : ||x_i - x_j|| = u\}$, i.e. the set of all pairs of data-points whose locations are a distance u apart, and $|N(u)|$ is the number of such pairs.

The variogram has long been a standard tool in classical geostatistical analysis; see, for example, Chilès & Delfiner (2016), who notes that it was be used to analyse data from Scandinavian forest surveys as early as 1926 (Langsaeter, 1926).

When the data locations x_i are irregularly distributed over the study area, typically the distances u_{ij} between pairs of locations are unique, hence $|N(u_{ij})| = 1$. The resulting empirical variogram, sometimes called the *variogram cloud*, is very unstable and difficult to interpret. Our recommended strategy is therefore to pool the data within pre-specified classes of distance, referred to as *bins*, and to average the empirical variogram within each of these classes.

Figure 2.8 shows this strategy in action for the river blindness and *A. gambiae* mosquito abundance data-sets. The left-hand panels show no obvious spatial structure, whereas in the right-hand panels binning with intervals of width 25 km (upper panel) and 10 km (lower panel) has revealed, for the river blindness data a rising trend up to a distance of around 150km followed by a drop, and for the *A. gambiae* data an alternating rise-and-fall pattern.

To establish whether the apparent patterns in the right-hand panels of Figure 2.8 are or are not compatible with random fluctuations about a constant value, we use the following Monte Carlo strategy to simulate the behaviour of empirical variograms under the assumption of spatial independence.

1. Randomly permute the labelling of the $\hat{Z}_i$ while holding fixed the locations x_i.

2. Compute the empirical variogram (2.11) using the permuted $\hat{Z}_i$.

3. Repeat 1 and 2 B times.

4. Use the resulting B empirical variograms to compute pointwise 95% tolerance intervals at each of the pre-specified distance bins under the hypothesis of spatial independence, i.e. a constant expected variogram.

If the empirical variogram lies outside outside the 95% tolerance band obtained in step 4, this is indicative of residual spatial correlation.

To formally test the hypothesis of spatial independence, we need to define a test statistic T, for example

$$T = \sum_{k=1}^{K} |N(u_k)| \left[\hat{V}(u_k) - \tau^2 \right]^2, \qquad (2.12)$$

where K is the number of distance bins.

Since τ^2 is almost always unknown, we can plug-in its maximum likelihood estimate in the fitted GLMM in (2.6). The null distribution of T can then be obtained using the simulated simulated variograms from step 2 of the Monte Carlo algorithm described above. Let $T_{(h)}$ denote the h-th sample from the null distribution of T, for $h = 1, \ldots, B$. Since evidence against spatial independence arises from large values of T, the p-value of the test can be computed as

$$\frac{1}{B} \sum_{h=1}^{B} I(T_{(h)} > t)$$

where $I(a > b)$ takes value 1 if $a > b$ and 0 otherwise, and t is the value of the test statistic obtained from the data. Strictly this is an approximate

p-value, but it can be made as precise as we wish by increasing the number of simulations, B.

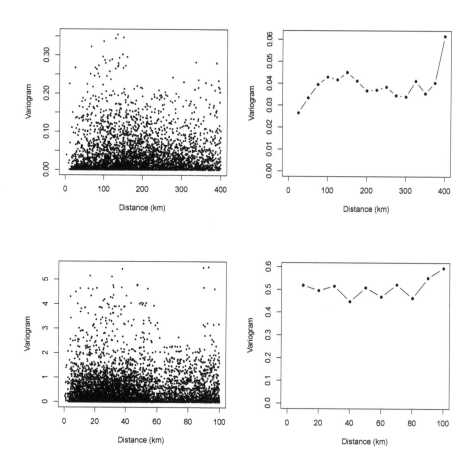

FIGURE 2.8
Empirical variograms of residuals from GLMM fits to data on river blindness in Liberia (upper panels) and abundance of *Anopheles gambiae* in Cameroon (lower panels). The left-hand panels shows un-binned empirical variograms, the right-hand panels empirical variograms using a bin-width of 25 km (upper panel) and 10 km (lower panel).

The grey shaded areas in each of the two left-hand panels of Figure 2.9 correspond to the 95% tolerance bands generated by the Monte Carlo algorithm, whilst the right-hand panels show the sampled values of the test statistic (2.12) with $B = 1000$ random permutations. The observed value for the unpermuted residuals $\hat{Z}_k$ is indicated by a solid dot on the abscissa and a vertical dotted line.

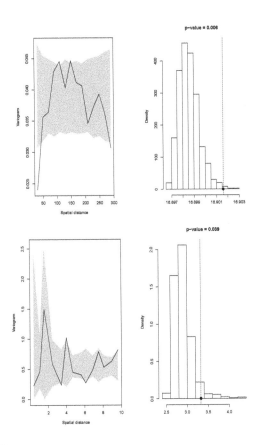

FIGURE 2.9
Plots of the diagnostic check on the presence of residual spatial correlation for the data on river blindness in Liberia (upper panels) and abundance of *Anopheles gambiae* in Cameroon (lower panels). Left panels: empirical variograms of the estimate residual variation $\hat{Z}_i$ (solid lines) and 95% tolerance bandwith (grey area) generated under the hypothesis of spatial independence. Right panels: null distribution of the test statistic in 2.12 with the solid point on the abscissa corresponding to the observed value of the test statistic.

In the case of the river blindness data, the results show clear evidence of residual spatial correlation. The empirical variogram lies partly outside the grey area, the p-value of the test based on (2.12) is 0.006 and, more importantly, the empirical variogram itself shows a rising trend up to a distance of about 100km. Note also that the falling trend in the empirical variogram at larger distances remains almost entirely within the grey shaded area. In our experience, this behaviour is typical. The empirical variogram can be calcuated for any distance up to the maxium distance between any two sampled locations, but its value as an exploratory tool is restricted to relatively small distances. As a rough rule of thumb, we would recommend restricting variogram calculations to distance up to about a quarter of the maximum linear dimension of the study region.

For the data on mosquito abundance, the evidence is much less strong. The observed variogram falls slightly outside the tolerance band for bins around 4 km and 10km. The test based on (2.12) yields a p-value of 0.039, but the empirical variogram shows no clear trend. Note that we have here restricted the variogram to distances no greater than 10km. This choice is consistent with our rule of thumb, but was also informed by inspection of the empirical variogram for these data shown in the lower right hand panel of Figure 2.8. There, we used 10km bins and observed a flat variogram. In the lower left-hand panel of Figure 2.9 we have drilled down to inspect the variogram over distance up to 10km to check whether the earlier choice of bins has hidden important small-scale spatial structure. The saw-tooth appearance of this empirical variogram is a warning that the numbers contributing to each bin are now uncomfortably small, resulting in highly variable estimates.

3

Theory

CONTENTS

3.1 Gaussian processes

A fundamental aim of any statistical modelling exercise is to describe, and ideally explain, the variation in a set of data. To achieve this aim, most models combine deterministic and stochastic elements. Perhaps the simplest example of this is the linear regression model, as described in Chapter 2. In a spatial context, this model can be expressed as

$$Y_i = \alpha + \beta d(x_i) + Z_i : i = 1, ..., n, \qquad (3.1)$$

where: Y_i is a measured outcome, or *response* at spatial location x_i; $d(x_i)$ is the value of an *explanatory variable* associated with x_i (deterministic variation); the Z_i are mutually independent random variables that, for present purposes, we can assume to be Normally distributed with mean zero and variance τ^2 (stochastic variation). Fitting the model to an observed data-set yields parameter estimates $\hat{\alpha}$, $\hat{\beta}$, $\hat{\tau}$, and *residuals*, $z_i = y_i - \hat{\alpha} - \hat{\beta}d(x_i)$. The residuals can be subjected to the usual diagnostic checks to decide whether the model provides an acceptable fit to the data. In a spatial context, these checks should include an investigation of the spatial distribution of the z_i, as described in Section 2.3.1. If these do show spatial structure, one explanation is that one or more spatially referenced explanatory variables other than $d(x)$ contribute to the variation in the response at location x. For example, with two additional explanatory variables $e(x)$ and $f(x)$, the model for the data becomes

$$Y_i = \alpha + \beta d(x_i) + \gamma e(x_i) + \delta f(x) + Z_i : i = 1, ..., n. \tag{3.2}$$

But in (3.2), what *are* these additional explanatory variables? If no candidates present themselves, we can express our lack of knowledge by considering their combined effects as a realisation of a stochastic process, $S(x)$. The model (3.2) then becomes

$$Y_i = \alpha + \beta d(x_i) + S(x_i) + Z_i : i = 1, ..., n, \tag{3.3}$$

in which the stochastic term $S(x_i)$ is a proxy for the combined effects of all unidentified, and therefore unmeasured, explanatory variables that influence the outcome at location x.

In (3.3), the critical distinction between the two stochastic elements $S(x_i)$ and Z_i is that the Z_i are mutually independent whereas the $S(x_i)$ are not. Specifically, the nature of the dependence between $S(x_i)$ and $S(x_j)$ is determined by their spatial locations; typically, we assume that pairs of values at close locations are more strongly dependent than pairs at distant locations (Tobler, 1970).

The number of ways that we could construct a stochastic model for a spatial surface $S(x)$ is unlimited. However, by far the most tractable, and for this reason most widely used, construction is a *Gaussian process*. The definition of a Gaussian process $S(x)$ is that for any finite set of locations $x_1, ..., x_n$, the joint probability distribution of $S(x_1), ..., S(x_n)$ is multivariate Normal. In our context, we can assume that the mean of each $S(x)$ is zero, because any non-zero mean is expressed through the regression component of (3.3). Since any multivariate Normal distribution is completely described by its mean vector and covariance matrix, it follows that to complete the specification of any particular Gaussian process we need only specify its *covariance function*, $\gamma(x, x') = \text{Cov}\{S(x), S(x')\}$, where x and x' are arbitrary locations. A convenient simplification is to assume that $S(x)$ is *stationary* and *isotropic*, whereby the variance of $S(x)$ is a constant, σ^2, and the correlation between $S(x)$ and $S(x')$ only depends on the distance between x and x', hence $\gamma(x, x') = \sigma^2 \rho(||x - x'||)$ where $|| \cdot ||$ denotes distance. Our remaining task is now to specify the *correlation function*, $\rho(u)$. Note that although a distance cannot be negative, $\rho(u)$ is always defined to be a symmetric function, i.e. $\rho(-u) = \rho(u)$, so as to be consistent with the fact that the covariance between two random variables must be symmetric in its arguments, i.e. $\gamma(x, x') = \gamma(x', x)$. We therefore only need to specify $\rho(u)$ for non-negative values of u.

Ensuring the validity of a particular choice for $\rho(u)$ is not a straightforward task. Consider, for example, the random variable

$$T = \int w(x)S(x)dx.$$

This has mean zero and variance

$$v = \sigma^2 \int\int w(x)w(x')\rho(||x - x'||)dxdx',$$

from which it follows that v must be non-negative, whatever choice we make for the function $w(x)$. This places severe, and non-obvious, constraints on allowable choices for the correlation function $\rho(u)$.

M. Stein (1999) gives a rigorous theoretical account of these issues. A practical recommendation is to choose from a catalogue of parametric families of correlation function that are known to be valid.

3.2 Families of spatial correlation functions

What general properties might we expect a family of spatial correlation functions to possess? Firstly, if we believe Tobler's first law of geography we might expect that the correlation between $S(x)$ and $S(x')$ should not increase as the distance between x and x' increases, i.e. $\rho(u)$ should be a non-increasing function of u. Secondly, since distances can be measured in a variety of units, $\rho(u)$ must include a scaling parameter

3.2.1 The exponential family

One valid correlation function that meets these requirements, is the *exponential correlation function*,

$$\rho(u; \phi) = \exp(-u/\phi) : u \geq 0, \tag{3.4}$$

where $\phi > 0$ is a single parameter. Figure 3.1 shows $\rho(u; \phi)$ for $\phi = 0.1$ and $\phi = 1.0$. In (3.4), the scaling parameter ϕ is also called the *range parameter*, because for a given unit of distance, a larger value of ϕ indicates that spatial correlation persists over a longer range. A conventional definition of the *practical range* of the spatial correlation is the distance at which $\rho(u) = 0.05$. For the exponential correlation function (3.4), the practical range is approximately three times ϕ (see Figure 3.1).

To convey an intuitive understanding of how the mathematical form of $\rho(u)$ translates into the physical properties of the underlying process, it is helpful to inspect simulated realisations of $S(x)$. Figure 3.2 gives an example of a realisation along a one-dimensional transect of length 10 for each of

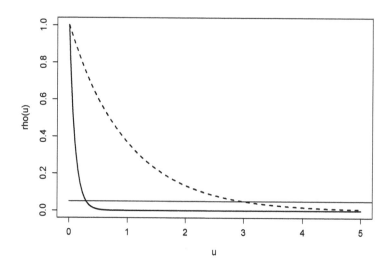

FIGURE 3.1
Exponential correlation functions, $\rho(u) = \exp(-u/\phi)$, for $\phi = 0.1$ and $\phi = 1.0$ (solid and dashed lines, respectively). The thin horizontal line intersects each $\rho(u)$ at its practical range, $u \approx 3 \times \phi$.

the two exponential correlation functions shown in Figure 3.1. In both cases, the process $S(x)$ has mean zero and variance 1. The most obvious difference between the two realisations, which both use the same underlying random number sequence, is that the realisation with $\phi = 1.0$ takes longer excursions away from its mean of zero than does the realisation with $\phi = 0.1$.

3.2.2   The Matérn family

The parameter ϕ gives us the flexibility to control the rate at which spatial correlations decay towards zero with increasing spatial separation. The Matérn family adds flexibility to the shape of the correlation function. This in turn has a direct bearing on the smoothness of the underlying spatial process.

Bertil Matérn (1917-2007) was a Swedish statistician who worked in the Royal College of Forestry, Stockholm from 1945 until his retirement in 1978. His PhD thesis (Matérn, 1960, reprinted as Matérn, 1986) had a profound influence on the later development of spatial statistical methods. One of its many contributions was the two-parameter family of correlation functions that bears his name[1]

---

[1]This may be a rare counter-example to Stigler's law of false eponymy (Stigler, 1980), which states that things are named after people who did not invent them.

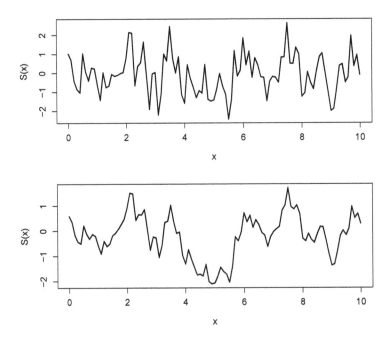

FIGURE 3.2
Realisations of stationary processes $S(x)$ with correlation function $\rho(u) = \exp(-u/\phi)$. The upper and lower panels correspond to $\phi = 0.1$ and $\phi = 1.0$, respectively.

The Matérn correlation function is

$$\rho(u; \phi, \kappa) = \{2^{\kappa-1}\Gamma(\kappa)\}^{-1}(u/\phi)^\kappa K_\kappa(u/\phi), \qquad (3.5)$$

where $\phi > 0$ and $\kappa > 0$ are parameters and $K_\kappa(\cdot)$ is the modified Bessel function of the third kind of order κ. Figure 3.3 shows $\rho(u; \phi, \kappa)$ for $\kappa = 0.5, 1.5, 2.5)$, in each case with ϕ chosen so that the practical range is 3.

The rather daunting algebraic expression (3.5) for the Matérn correlation function conceals an elegant interpretation. If $\kappa > d$, the Matérn correlation is d times differentiable at $u = 0$, and the underlying Gaussian process $S(x)$ is then also d times differentiable (strictly, differentiable in mean square, see M. Stein (1999)). This matters because, as we shall see in later chapters, if we model a set of data using the Matérn correlation function, predictive inferences inherit the differentiability of $\rho(u; \phi, \kappa)$. The exponential correlation function is the special case of the Matérn when $\kappa = 0.5$; hence, an exponentially correlated Gaussian process $S(x)$ is mean-square continuous but non-differentiable. Figure 3.4 shows a realisation of each of the corresponding Gaussian processes

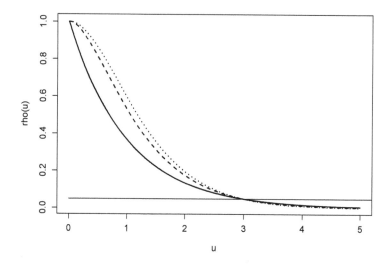

FIGURE 3.3
Matérn correlation functions, $\rho(u; \phi, \kappa)$, for $\kappa = 0.5, 1.5, 2.5$ (solid, dashed and dotted lines, respectively) and values of ϕ chosen so that in each case the practical range is 3.0. The thin horizontal line at height 0.05 therefore intersects each correlation function at the same value, $u = 3$.

$S(x)$ along a one-dimensional transect for $\kappa = 1.5$ and $\kappa = 2.5$, corresponding to differentiable and twice-differentiable process $S(x)$; in both cases, the practical range is 3.

3.2.3 The spherical family

Our experience has been that the Matérn family is sufficiently flexible for most applications. One pragmatic reason for this is that even within the restriction to the Matérn family, a surprisingly large amount of data is needed in order to estimate both ϕ and κ, and it is therefore reasonable to *choose* the value of κ from a limited set of values, for example $\kappa = 0.5, 1.5, 2.5$ as in Figures 3.3 and 3.4. Nevertheless, other parametric families of correlation function have been proposed, and some of these are widely used in classical geostatistics. Here, we describe only one; for a more detailed discussion, see Chilès & Delfiner (2016).

The spherical correlation function is the one-parameter family given by

$$\rho(u) = \begin{cases} 1 - \frac{3}{2}(u/\phi) + \frac{1}{2}(u/\phi)^3 & : u \le \phi \\ 0 & : u > \phi \end{cases} \tag{3.6}$$

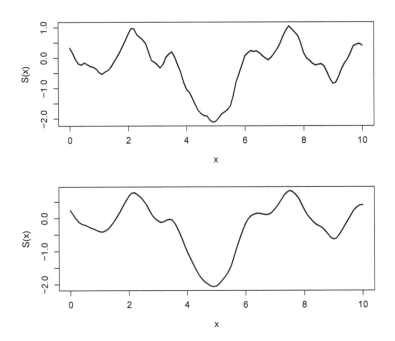

FIGURE 3.4
Realisations of stationary processes $S(x)$ with Matérn correlation function $\rho(u; \phi, \kappa)$ The upper and lower panels correspond to $\kappa = 1.5$ and $\kappa = 2.5$, respectively.

This corresponds to a continuous but non-differentiable process $S(x)$ whose correlation function is exactly zero at all distances greater than ϕ. Figure 3.5 compares spherical and exponential correlation functions with values of ϕ chosen so that both have the same practical range, $u = 3$. The spherical would better capture an approximately linear rising trend in an empirical variogram but, echoing our earlier comment about the difficulty of estimating both ϕ and κ for the Matérn, to differentiate unequivocally between models that assume exponential and spherical correlations might require a substantial amount of data.

3.2.4 The theoretical variogram and the nugget variance

In Chapter 2 we introduced the *empirical variogram* as a useful device for diagnostic checking of the residuals from a linear regression analysis. The

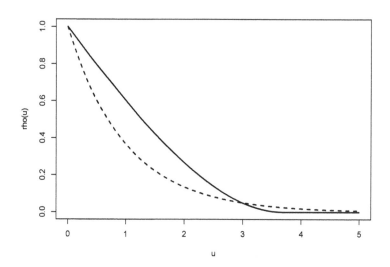

FIGURE 3.5
Spherical correlation functon and Matérn correlation function with $\kappa = 1.5$
(solid and dashed lines, respectively) with values of ϕ chosen so that in each
case the practical range is 3.0.

theoretical counterpart of this is the function

$$V(x, x') = \frac{1}{2}\mathrm{Var}\{S(x) - S(x')\},\tag{3.7}$$

where x and x' are any two points in $\mathbb{R}^2$. If the process $S(x)$ is stationary,
with variance σ^2 and correlation function $\rho(u)$, it follows that

$$
\begin{aligned}
V_S(x, x') &= \frac{1}{2}(\mathrm{Var}\{S(x)\} + \mathrm{Var}\{S(x')\} - 2\mathrm{Cov}\{S(x), S(x')\}) \\
&= \sigma^2\{1 - \rho(u)\}.
\end{aligned}\tag{3.8}
$$

This, incidentally, explains why (3.7) includes the factor $\frac{1}{2}$, and why some
authors call $V(u)$ the *semi-variogram*.

When we observe a geostatistical data-set, $(x_i, Y_i) : i = 1, ..., n$, we can
easily conceive of a spatially continuous process $Y(x)$, but if the physical
process by which we measure each Y_i is imprecise, then the value of $Y(x)$
is not unique; if we take two measurements Y_1 and Y_2 at the same location
x, they may well have different values. A tangible example is when a blood-
sample is split into two sub-samples, each of which is assayed separately to
determine its biochemical composition.

Mathematically, this phenomenon can be represented by specifying the correlation function of $Y(x)$ to be discontinuous at $u = 0$, i.e. the correlation between $Y(x)$ and $Y(x')$ approaches a value less than 1 as the distance between x and x' approaches zero. We represent this by writing the data-value Y_i as

$$Y_i = S(x_i) + Z_i : i = 1, ..., n \qquad (3.9)$$

where the Z_i are mutually independent with zero mean and variance τ^2. If, in (3.9), $S(x)$ has variance σ^2 and correlation function $\rho(u)$, where u denotes the distance between the data-locations x_i and x_j, then $\text{Corr}\{Y_i, Y_j\} = \sigma^2 \rho(u)/(\tau^2 + \sigma^2)$, which approaches $\sigma^2/(\tau^2 + \sigma^2)$ as u approaches zero. The equivalent expression as a variogram is

$$V(u) = \tau^2 + \sigma^2\{1 - \rho(u)\}. \qquad (3.10)$$

Comparing (3.8) and (3.10), we see that the effect of measurement error in our data-values Y_i is to add a positive constant to the theoretical variogam.

The measurement error variance τ^2 is also called the *nugget variance* or simply *nugget*, a name that recalls the origins of classical geostatistics in the South African mining industry. Figure 3.6 shows a generic variogram with its key properties highlighted. In addition to the terms *nugget* and *practical range* introduced earlier, the variance of $S(x)$ is sometimes called the *sill*.

This nomenclature also reminds us that a nugget need not be a measurement error. Suppose, for example, that our data were generated by the model

$$Y_i = S(x_i) + W(x_i) + Z_i : i = 1, ..., n \qquad (3.11)$$

where, in addition to previously defined terms, $W(x)$ is a spatial process with variance ν^2 and correlation function $\rho_W(u)$ with the property that $\rho_W(u) = 0$ for all u greater than a distance d, say. If we now imagine that the data-locations x_i form a regular square lattice with spacing greater than d, the correlation between any two $W(x_i)$ and $W(x_j)$ is zero and the augmented model (3.11) is indistinguishable from the simpler model

$$Y_i = S(x_i) + Z_i^* : i = 1, ...,,$$

where the Z_i^* are mutually independent with variance $\nu^2 + \tau^2$.

In chapter 6 we will consider the implications of the above discussion for the design of geostatistical investigations. For the time being, we note that in many applications an estimated nugget variance represents a combination of measurement error and small-scale spatial variation, in unknown proportions.

3.3 Statistical inference

Statistical inference is the process of drawing formal conclusions from data.

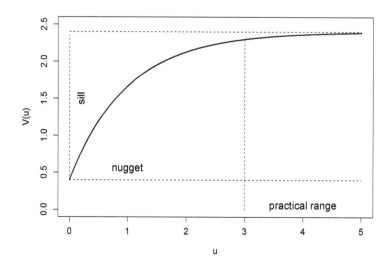

FIGURE 3.6
The generic form of the variogram, $V(u)$. The *nugget* is another name for the
parameter τ^2, the *sill* is likewise another name for the parameter σ^2, whilst the
practical range is the distance u at which the correlation function $\rho(u)$ decays
to 0.05.

Particular forms of inference include testing, estimation and prediction. In the
remainder of this chapter we describe only those specific aspects of inference
that we shall use in the later chapters. More detailed discussions of likelihood-
based inference can be found in many textbooks, including Azzalini (1996)
and Pawitan (2001). Books that give a more detailed discussion of Bayesian
inference include O'Hagan (1994) and Lee (2012).

3.3.1   Likelihood-based inference

A statistical model specifies the joint probability distribution for a set of
data. A generic notation for this is $[Y; \theta]$, to be read as "the distribution of Y
given θ," where Y denotes the data and θ a set of unknown constants, called
parameters. Strictly, Y is the multivariate random variable of which the data,
y, are a realisation. Using upper and lower case letters to distinguish a random
variable from its observed value is a useful convention.

    Broadly speaking, statistical *testing* addresses questions like "are the ob-
served data compatible with a pre-specified, or *hypothesised* value of θ?"
whereas statistical *estimation* addresses questions like "with what range of

values of θ are the observed data compatible?" It follows that testing is only useful when the hypothesised value of θ has a particular status.

Statistical *prediction* involves making a probability statement about an unobserved random variable. In the simplest geostatistical problem, the data Y are the realised values $Y_i = S(x_i)$ at locations $x_i : i = 1, ..., n$ and we want to say something about the realised, but unobserved value of $S(x)$ at an arbitrary location x. To include problems of this kind we need to extend our notation to

$$[Y; \theta] = \int [Y|S; \theta][S; \theta]dS. \qquad (3.12)$$

To understand (3.12), note firstly that $[S; \theta]$ is, as before, to be read as "the distribution of S given θ" where S is to be understood as the values of $S(x)$ at *all* locations x in the geographical region of interest. Secondly, $[Y|S; \theta]$ is to be read as "the distribution of Y given S and θ." Here, the vertical bar denotes conditioning on the value of an unobserved random variable, whereas the semi-colon denotes conditioning on the value of an unknown constant. This distinction may seem pedantic, and indeed some statisticians consider it unnecessary (see, for example, Section 3.4 below), but we think it important. For example, if a scientist chooses to investigate a spatial phenomenon $S(x)$ in several different geographical regions, and is prepared to assume that the underlying natural processes that generate $S(x)$ are the same in all of them, then the value of θ will be the same in all regions, but the realisations of $S(x)$ will be different. Finally, the integral expresses the rule of elementary probability that the marginal distribution of Y is obtained by integrating the joint distribution of Y and S with respect to S.

A useful terminology, suggested to us by Noel Cressie, is to refer to the distribution $[S; \theta]$ in (3.12) as the *process model* and the distribution $[Y|S; \theta]$ as the *data model*; see Figure 3.7. Statistical models that incorporate unobserved random variables are also called *hierarchical models*. Generic names for unobserved random variables include *latent variables* or *random effects*.

The *likelihood* function associated with a statistical model is the joint probability distribution of the data considered as a function of the parameters, with the data held fixed at their observed values. For technical reasons, the likelihood is more usefully expressed on a log-transformed scale, hence the *log-likelihood* is the function

$$L(\theta) = \log[y; \theta]. \qquad (3.13)$$

The log-likelihood function is the cornerstone of modern statistical inference. It provides a completely general approach to testing and estimation that usually leads to operational procedures with good theoretical properties. Two key results are the following.

R1. *Maximum likelihood estimation*

The maximum likelihood estimate of θ is the value that maximises $L(\theta)$, written as $\hat{\theta}$. In large samples, the probability distribution of $\hat{\theta}$ is multivariate

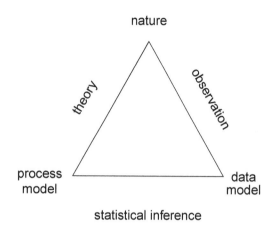

FIGURE 3.7
A process model expresses a scientist's assumptions about nature, a data model expresses a statistician's assumptions about how nature generates observations, statistical inference links the two.

Normal, with mean θ and variance matrix that can be approximated by $V(\theta) = -\{L''(\hat{\theta})\}^{-1}$, where $L''(\theta)$ denotes the matrix of second derivatives whose $(i,j)^{th}$ element is

$$\frac{\partial^2 L(\theta)}{\partial\theta_i\theta_j}.$$

The practical implication of this result is that, to a good approximation, maximum likelihood estimates are unbiased, with known variance matrix. The result can be used to calculate approximate confidence intervals for individual parameters θ_i by the usual formula, $\hat{\theta}_i \pm 1.96\sqrt{\{\mathrm{Var}(\hat{\theta}_i)\}}$. It can also be used to carry out a *Wald test* of a hypothesised value of θ_i, by comparing the Z-statistic, $(\hat{\theta}_i - \theta_i)/\sqrt{\{\mathrm{Var}(\hat{\theta}_i)\}}$ with critical values of the standard Normal distribution.

R2. *Deviance tests*

The second key result is that if the parameter vector θ has q elements, then the *deviance function*, $D(\theta) = 2\{L(\hat{\theta}) - L(\theta)\}$ has a chi-squared distribution on q degrees of freedom. This result can be used directly to test a hypothesised value of the complete parameter vector θ, and indirectly to calculate a joint confidence region for θ. If c denotes the p-quantile of χ_q^2, i.e the value such that $P\{\chi_q^2 > c\} = p$, then a $100(1-p)\%$ *likelihood-based confidence region* for θ is the set of all values of θ for which $D(\theta) < c$.

Calculating a likelihood-based confidence region is straightforward in principle but difficult in practice, rapidly becoming impossible as the number of elements of θ grows. However, a more general version of the result also holds, and is extremely useful in practice.

Suppose that θ has dimension $q + r$, and denote this by writing $\theta = (\theta_1, \theta_2)$ where θ_1 and θ_2 have dimension q and r, respectively. then, $D(\theta) = 2\{L(\hat{\theta}_1, \hat{\theta}_2) - L(\hat{\theta}_1, \theta_2)\}$ again has a chi-squared distribution on q degrees of freedom. This more general result allows us to test hypotheses about, and to construct confidence regions for, arbitrary sub-sets of θ, including single elements. The results will not be identical to Wald tests or confidence intervals calculated using R1, but they will be approximately so in large samples.

One of the commonest applications of R2 is to compare *nested* models, meaning that one is a special case of the other, typically obtained by setting one or more parameters of the larger model to zero.

The results R1 and R2 are invoked routinely in modern statistical practice, but they rely on assumptions that are not always satisfied. In a geostatistical context, the two most common situations in which these assumptions do not hold are *boundary problems* and *dimensional ambiguities*. We can illustrate both of these using the geostatistical model (3.9), which we reproduce here as

$$Y_i = S(x_i) + Z_i : i = 1, ..., n$$

Recall that $S(x)$ is a Gaussian process with mean zero, variance σ^2 and correlation function that, for illustration, we assume is $\rho(u) = \exp(-u/\phi)$, whilst the Z_i are mutually independent $N(0, \tau^2)$. The model therefore has three parameters, $\theta = (\sigma^2, \phi, \tau^2)$. Suppose we wish to test the hypothesis that $\tau^2 = 0$, i.e. the Y_i are measured without error. Since τ^2 is a variance, it cannot be negative, i.e. the hypothesised value $\tau^2 = 0$ lies on the boundary of the parameter space. It turns out that there is a neat fix to this problem due to Self & Liang (1987), namely to replace the nominal significance level p by $p/2$. Whether it makes scientific sense to test this hypothesis is another matter,either because we know from the context of the problem that the Y_i are not measured without error or, as we have shown above, even when the data strictly contain no measurement error, the Z_i term in the model can act as a proxy for small-scale spatial variation.

At first glance, testing whether there is any spatially correlated variation at all, i.e. hypothesising that $\sigma^2 = 0$, is a second example of a boundary problem. However, we could also define absence of spatial correlation as $\phi = 0$. If $\sigma^2 = 0$, the value of ϕ becomes irrelevant and *vice versa*. Put another way, does the hypothesis of no spatially correlated variation reduce the number parameters from three to two, or to one? This is an example of dimensional ambiguity, and in assessing this hypothesis using the deviance, we would need to know which, if either, of the χ_1^2 or χ_2^2 distributions would give the correct critical value c.

Fortunately, and as already hinted, there are good scientific reasons why, in the remainder of this book we will not be advocating formal testing of variance components. But it would be wrong to leave the reader with the idea that the results R1 and R2 are universally, as opposed to very widely, applicable.

3.4 Bayesian Inference

In likelihood-based inference, parameters are considered to be constants, whose values are unknown and cannot be measured directly (the literal translation of the ancient Greek "parameter" is "beyond measurement"). In Bayesian inference, parameters are considered to be unobserved (and unobservable) random variables. It follows that parameters θ and data Y must be assigned a joint probability distribution, which we denote by $[Y, \theta]$ and which can be factorised as the product of the marginal distribution of θ and the conditional distribution of Y given θ. We write this as

$$[Y, \theta] = [Y|\theta][\theta]. \tag{3.14}$$

The expression $[Y|\theta]$ in (3.14) is algebraically identical to the expression $[Y; \theta]$ introduced at the beginning of Section 3.3.1. However, the philosophical change in the status of θ, from an unknown constant to an unobserved random variable, has far-reaching implications for how we approach statistical inference. The first of these is that we can equally well factorise the joint distribution of Y and θ as

$$[Y, \theta] = [Y][\theta|Y],$$

which we can re-arrange to give

$$[\theta|Y] = [Y, \theta]/[Y] = [Y|\theta][\theta]/[Y].$$

The marginal distribution of Y can also be written as

$$[Y] = \int [Y, \theta] d\theta = \int [Y|\theta][\theta] d\theta.$$

Combining the last two equations then gives the equation

$$[\theta|Y] = [Y|\theta][\theta] / \int [Y|\theta][\theta] d\theta. \tag{3.15}$$

Equation (3.15) is called *Bayes' Theorem* after the Reverend Thomas Bayes (1702-1761)[2].

---

[2]Who may or may not have discovered it – recall Stigler's Law

So far, so just maths: the algebraic manipulations leading to (3.15) are simply an application of elementary probability theory. However, (3.15) is also the foundation for a different approach to statistical inference, in which the marginal distribution $[\theta]$, called the *prior distribution for θ*, reflects uncertainty about $[\theta]$ before collecting the data Y, whilst the conditional distribution $[\theta|Y]$, called the *posterior distribution for θ*, reflects uncertainty about θ after collecting the data.

The likelihood function, $[Y;\theta]$ or $[Y|\theta]$ according to taste, plays a central role in both forms of inference. The crucial additional requirement for Bayesian inference is specification of the prior distribution, $[\theta]$. One view of this is that $[\theta]$ expresses an investigator's prior belief about θ, in which case it is perfectly acceptable for two scientists (or statisticians) to agree on the statistical model, $[Y|\theta]$, but to specify different priors and therefore reach different conclusions from the same data. Another approach, called *prior elicitation*, tries to reach a consensus from subject-matter experts on what prior should be used. A more pragmatic approach is to argue that conclusions should not be greatly influenced by the choice of prior; roughly speaking, this is achieved by specifying a *diffuse* prior; an example of a diffuse prior for a parameter that could in principle take any real value would be a Normal distribution with mean zero and standard deviation an order of magnitude bigger than any sensible range of values that θ might take in context.

In Bayesian inference, the closest formal counterpart to a hypothesis test might be to use a prior distribution that assigns a positive probability to the hypothesised value of θ. This begs the question of what value this probability should take, and we are not aware of it ever being used in practice. Bayesian concepts of parameter estimation are more directly comparable with their likelihood-based counterparts. A Bayesian point estimator for an element of θ is the value at which the posterior distribution is maximised. A Bayesian *credible interval* is defined as a range of values over which the posterior distribution integrates to a specified value, conventionally 0.95.

3.5 Predictive inference

A formal definition of *statistical prediction* is the following. To simplify the description we will temporarily ignore the role played by parameters, θ.

Let $[Y]$ denote a probability model for a set of data, and write y for the observed value of Y. Suppose we want to use these data to say something about the realised value of an unobserved random variable T that is stochastically dependent on Y. We call T the predictive *target*, or *predictand*, and define the *predictive distribution* of T to be the conditional distribution, $[T|Y = y]$. A *point prediction* of T can then be calculated using a summary measure of its predictive distribution, for example the mean, median or mode. One argument for using the mean is the following general result.

Theorem 3.5.1. Let $\tilde{T} = t(Y)$ be any function of Y and define the *mean square error* of $\tilde{T}$ to be the quantity $\mathrm{MSE}(\tilde{T}) = \mathrm{E}[\{T - t(Y)\}^2]$. Then, $\mathrm{MSE}(\tilde{T})$ takes its smallest possible value when $t(Y) = \mathrm{E}[T|Y]$.

Proof. First, write $\mathrm{MSE}(\tilde{T}) = \mathrm{E}_Y[\mathrm{E}_T[\{T - t(Y)\}^2|Y]]$. Next, write the inner expectation as

$$
\begin{aligned}
\mathrm{E}_T[\{T - t(Y)\}^2|Y] &= (\mathrm{E}_T[\{T - t(Y)\}|Y])^2 + \mathrm{Var}[\{T - t(Y)\}|Y] \\
&= (\mathrm{E}_T[T|Y] - t(Y))^2 + \mathrm{Var}(T|Y) \quad\quad (3.16)
\end{aligned}
$$

where, in (3.16) we have used the fact that, conditional on Y, any function of Y is a constant.

Now, on the right hand side of (3.16) the second term, $\mathrm{Var}(T|Y)$ does not depend on the function $t(\cdot)$, whilat the first, $(\mathrm{E}_T[T|Y] - t(Y))^2$, is non-negative and equal to zero if and only if $t(Y) = \mathrm{E}_T[T|Y]$. This completes the proof and shows, incidentally, that the achieved mean square error is $\mathrm{E}_Y[\mathrm{Var}(T|Y)]$.

Theorem 3.5.1 is not specific to geostatistics, but we shall use it in Chapter 4 to provide a model-based justification for the very widely used geostatistical prediction method known as *kriging*.

In reality, any useful geostatistical model will involve unknown parameters, θ. One way to deal with this is by using the *plug-in* predictive distribution, in which θ is replaced by an estimate, $\hat{\theta}$. Specifically, writing the predictive distribution of T as $[T|Y; \theta]$ to acknowledge its dependence on unknown parameters, the plug-in predictive distribution is $[T|Y; \hat{\theta}]$, and in likelihood-based inference the estimate of choice is the maximum likelihood estimate.

A legitimate concern about plug-in prediction is that it does not take into account the uncertainty in $\hat{\theta}$. Bayesian inference provides an elegant solution to this problem. Since θ is now a random variable, it follows that

$$
[T|Y] = \int [T, \theta|Y]d\theta = \int [T|Y, \theta][\theta|Y]d\theta. \quad\quad (3.17)
$$

Equation (3.17) shows that the Bayesian predictive distribution is a weighted average of plug-in predictive distributions, with weights attached to different values of θ according to their posterior probabilities. A likelihood-based alternative is to use the asymptotic multivariate Normal sampling distribution of $\hat{\theta}$ in place of the Bayesian posterior $[\theta|Y]$ as the weighting distribution in (3.17).

3.6 Approximations to Gaussian processes

The computations associated with geostatistical estimation and prediction can be burdensome for large data-sets. Various approximate methods are available for easing this computational burden. Here, we describe only the two that we will use in later chapters. Both methods meet the following two requirements. Firstly, they use only a finite number of random variables to define a spatially continuous process on a region of interest A . Secondly, their properties mimic those of the spatially continuoius Gaussian process models that we described in Sections 3.1 and 3.2.

This section necessarily contains some highly technical material. Readers whose main interest is in applications may wish to regard it as an optional extra.

3.6.1 Low-rank approximations

To motivate the idea of a low-rank approximation, we digress briefly to describe the *kernel regression* method for estimating a function $s(x)$ using data $Y_i = s(x_i) + Z_i$, where the Z_i are mutually independent measurement errors. If we specify $s(x)$ as a parametric family of functions, for example a polynomial in the coordinates of x, this is a standard regression problem that can be solved using the method of least squares. If all we are willing to assume is that $s(x)$ varies smoothly with x, we could instead estimate $s(x)$ as a weighted average of the data,

$$\hat{s}(x) = \sum_{i=1}^{n} w_i(x)Y_i / \sum_{i=1}^{n} w_i(x), \qquad (3.18)$$

placing relatively large weights on measured values Y_i at locations x_i close to x. A *kernel regression* estimate does precisely this, defining the weights $w_i(x)$ in (3.18) as follows. Choose a *kernel function* $f(u)$, a non-negative-valued function of u, symmetric about $u = 0$ and with a single mode at $u = 0$. Then,

$$w_i(x) = f\{(x - x_i)/\phi\}. \qquad (3.19)$$

Figure 3.8 shows this process in one dimension, highlighting how the character of the kernel estimate $\hat{s}(x)$ changes as ϕ increases. The kernel function is a truncated quartic

$$f(x) = \begin{cases} (1 - u^2)^2 & -1 \le u \le 1 \\ 0 & \text{otherwise} \end{cases}$$

All of the estimates $\hat{s}(x)$ shown in Figure 3.8 are mathematically smooth in the sense of being twice differentiable everywhere, but in an everyday sense the larger the value of ϕ the smoother the appearance of the estimate in the sense that it oscillates less rapidly.

The intuitive idea behind a low-rank approximation to $S(x)$ is to apply the kernel regression estimate not to a set of observed values Y_i but to a set of unobserved, mutually independent random variables located at the points

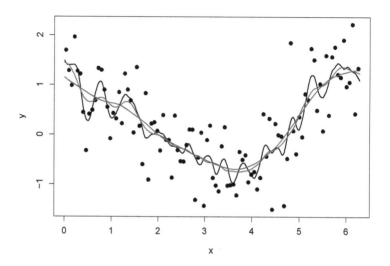

FIGURE 3.8
Kernel regression estimates of a function $s(x)$ calculated from data Y_i observed at 100 equally spaced points x_i in one spatial dimension. Black, red and blue lines correspond to values of $\phi = 0.2, 0.4$ and 0.8, respectively.

of a regular grid that covers the region of interest, A. Hence,

$$S_m(x) = m^{-0.5} \sum_{k=1}^{m} U_k \times (\phi^{-1} f\{(x - x_k)/\phi\}), \qquad (3.20)$$

where the U_k are mutually independent $N(0, \sigma^2)$. The scaling by $m^{-0.5}$ on the right hand side of (3.20) looks arbitrary, but the reason for it will become clear as we explore the properties of the resulting process $S_m(x)$. Note also that $S_m(x)$ is unchanged if we multiply σ and divide $f(\cdot)$ by the same amount. To avoid this ambiguity, we scale the kernel function so that $f(x)$ integrates to 1 over $\mathbb{R}^2$, in other words $f(x)$ is a probability density on $\mathbb{R}^2$.

Our use of σ^2 and ϕ for the parameters of the model defined by (3.20) echos our use of the same terminology for the variance and correlation range parameters of the models considered in Sections 3.1 and 3.2. This is not coincidental. Figure 3.8 hints, and it turns out to be true, that σ^2 and ϕ here are directly related to the variance and correlation range of $S_m(x)$. Also, as we now show, our choice of the scaling factor in the definition of $S_m(x)$ means that the covariance structure of (3.20) has a well-defined limiting form as the lattice spacing of the points x_k approaches zero, and (3.20) can therefore be interpreted as a numerical approximation to a spatially continuous process, $S^*(x)$ say, of the kind described in Sections 3.1 and 3.2.

Firstly, note that the variance of $S_m(x)$ is given by

$$\phi^{-2}m^{-1}\mathrm{E}\left[\left\{\sum_{j=1}^{m}U_j f((x-x_j)/\phi)\right\}\left\{\sum_{k=1}^{m}U_k f((x-x_k)/\phi)\right\}\right]$$

Since $\mathrm{E}[U_j^2] = \sigma^2$ and $\mathrm{E}[U_j U_k] = 0$ for all $j \neq k$, this reduces to

$$v(x) = \phi^{-2}\sigma^2 m^{-1}\sum_{j=1}^{m}f((x-x_j)/\phi)^2. \tag{3.21}$$

More generally, the covariance between $S_m(x)$ and $S_m(x')$ is given by

$$g(x,x') = \phi^{-2}\sigma^2 m^{-1}\sum_{j=1}^{m}f((x-x_j)/\phi)f((x'-x_j)/\phi). \tag{3.22}$$

Now, for any region A, if h denotes the spacing of the m lattice points that cover A, then $m \propto 1/h^2$, and as h approaches zero,

$$m^{-1}\sum_{j=1}^{m}f((x-x_j)/\phi)^2 \propto h^2\sum_{j=1}^{m}f((x-x_j)/\phi)^2 \to \int_A f((x-y)/\phi)^2 dy.$$

This holds for any A, and as A increases without limit to approach the entire plane, $\mathbb{R}^2$, the integral converges to

$$\int_{\mathbb{R}^2} f(y/\phi)^2 dy,$$

whatever the value of x. Similarly,

$$g(x,x') \to \phi^{-2}\sigma^2 \int_{\mathbb{R}^2} f(y/\phi)f((y-u)/\phi)dy,$$

where $u = x - x'$. This gives the limiting form of (3.21) as a Gaussian process $S(x)$ with covariance function $\sigma^2 g(u;\phi)$ where

$$g(u;\phi) = \phi^{-2}\int_{\mathbb{R}^2} f(y/\phi)f((y-u)/\phi)dy. \tag{3.23}$$

The variance and correlation function follow as $v = \sigma^2 g(0;\phi)$ and $\rho(u;\phi) = g(u;\phi)/g(0;\phi)$, respectively. It is not immediately obvious from the expression on the right-hand side of (3.23) that v does not depend on the value of ϕ. To see this, we set $u = 0$ in (3.23) and evaluate the integral using polar coordinates (r, θ) to give

$$\begin{aligned}
v &= \sigma^2\phi^{-2}\int_0^{2\pi}\int_0^{\infty} f(r/\phi)^2 v\, dr\, d\theta \\
&= 2\pi\sigma^2\int_0^{\infty} f(v)^2 v\, dv. \tag{3.24}
\end{aligned}$$

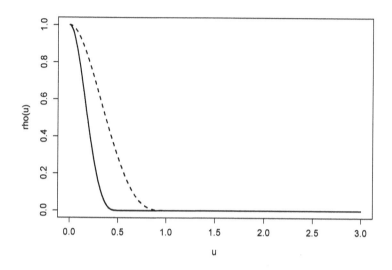

FIGURE 3.9

The correlation function corresponding to (3.23) with kernel function $f(u) = 2\pi^{-1}(1-||u||^2) : ||u|| \leq 1$ and $\phi = 1, 0.5$ (solid and dashed lines, respectively).

The form of the dependence of $\rho(u; \phi)$ on ϕ is also not immediately obvious, but in fact ϕ plays essentially the same role as does the correlation range parameter in the models described in Sections 3.1 and 3.2. Figure 3.9 illustrates this using a quadratic kernel,

$$f(u) = \begin{cases} 2\pi^{-1}(1 - ||u||^2) & : ||u|| \leq 1 \\ 0 & : \text{otherwise} \end{cases} \qquad (3.25)$$

Note that in this case, $\rho(u; \phi) = 0$ for $u > 2\phi$.

The correlation functions shown in Figure 3.9 are qualitatively similar to members of the Matérn family. To emphasise this, Figure 3.10 shows a Matérn correlation with $\phi = 0.2$ and $\kappa = 4$, together with the limiting form of the low-rank correlation when $\phi = 0.9$, a value chosen to make the latter a reasonable approximation to the former.

3.6.2 Gaussian Markov random field approximations via stochastic partial differential equations

Consider an outcome $Y(x_i)$, for $i = 1, \ldots, n$, measured over a regular grid $\{x_1, \ldots, x_n\}$. Also, use $\mathcal{N}_i$ to denote the set of neighbouring grid cells to x_i. First order neighbours (orange squares in Figure 3.11) to a given cell x_i are those cells that share a common edge with x_i. Second order neighbours (yellow

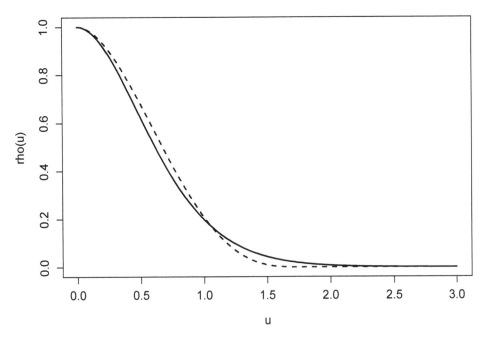

FIGURE 3.10
A Matérn correlation function (solid line) and the limiting form of the correlation function of an approximating low-rank model (3.23) with kernel function $f(u) = 2\pi^{-1}(1 - ||u||^2) : ||u|| \leq 1$ and $\phi = 0.9$ (dashed line).

and green squares in Figure 3.11) have, instead, a common edge with the first order neighbours, and so forth.

The measurements $Y(x_i)$ then follow a Gaussian Markov random field (GMRF) distribution if $Y(x_i)$ conditionally on all $Y(x_j)$ for $j \neq i$ (also referred to as the full conditional) only depends on on its neighbours $\{Y(x_j) : j \in \mathcal{N}_i\}$. However, in order to define a proper probability distribution, there are constraints on the forms that the full conditionals can take, which are defined by the Hammersely-Clifford theorem; we refer the interested reader to Besag (1974). The main appeal of GMRF models arises from their mathematical tractability and computational efficiency. Indeed, the precision matrix (i.e. the inverse of the covariance matrix), Q say, of a GMRF has (i, j)-th entry different from 0 if and only if x_i and x_j are neighbours. This result is a corollary to the more general result that any pair of Gaussian variables $Y(x_i)$ and $Y(x_j)$ are conditionally independent if and only if the (i, j)-th entry of Q is 0. When Q is a sparse matrix, meaning that most of its entries are zero, storage and manipulation of Q is easier and specialized algorithms that takes advantage of the sparse structure can then be developed.

Now consider a GMRF such that

$$Y(x_i)|\{Y(x_j) : i \sim j\} \sim N\left(\frac{1}{\delta}\sum_{j \in \mathcal{N}_i} Y(x_j), \frac{1}{\delta}\right)$$

where $|\delta| > 4$ and $\mathcal{N}_i$ is the set of first order neighbours. For this model, Besag (1981) has shown that

$$\text{Cov}\{Y(x_i), Y(x_j)\} \approx \frac{\delta}{2\pi}K_0(h\sqrt{\delta - 4}),$$

where $h = \|x_i - x_j\|$ is the Euclidean distance between x_i and x_j. The right hand side of the above equation is also obtained as a limiting case for the Matérn covariance function in (3.5), by setting $\phi^2 = (\delta - 4)^{-1}$ and $\sigma^2 = \delta/(4\pi)$, as $\kappa \to 0$ (henceforth, a Matérn zero process). Whittle (1953, 1963) shows in fact that a Matérn zero process $S(x)$ with scale ϕ is the solution to the following partial stochastic differential equation (i.e. a partial differential equation in which one or more of the terms is a stochastic process),

$$\phi^{-2} - \frac{\partial^2 S(x)}{\partial^2 a} - \frac{\partial^2 S(x)}{\partial^2 b} = W(x), x \in A \qquad (3.26)$$

where a and b are the abscissa and ordinate of x, and $W(x)$ is Gaussian white noise.

Equation (3.26) can then be used to approximate a Matérn zero process with a GMRF where the property of conditional independence is defined in terms of the first order neighbours. This GMRF is also called a first order conditional autoregressive (CAR) process. Intuitively, this suggests that a CAR process of order $\kappa + 1$ on a grid can be used to approximate a Matérn process with a positive integer smoothness parameter κ. Lindgren et al. (2011) have indeed proved this result, and use it to develop a general approach to approximating a Matérn process using CAR models, when κ is any positive real number. Figure 3.11 shows the entries of the precision matrix of the approximating CAR process for a given grid location x (red cell) when $\kappa = 1$ (left panel) and $\kappa = 2$ (right panel). Figure 3.12 shows the accuracy of the CAR approximation using a 64 by 64 regular grid for a Matérn process with $\phi = 0.1$ and $\kappa = 1$.

An apparent limitation to the Lidgren et al approximation is that geostatistical data arising from public health studies typically are not arranged on a regular grid but are irregularly scattered within the area of interest A. Hopwever, Lindgren et al. (2011) have tackled this issue by proposing to use a subdivision of A into non-intersecting triangles, where any two triangles meet in at most a common edge or vertex. They then define a CAR process over the vertices of the triangles. The resulting approximation of the Matérn process

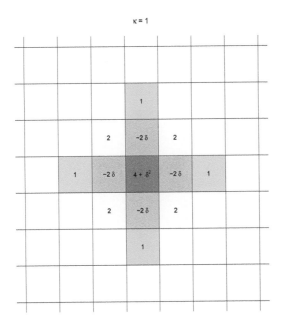

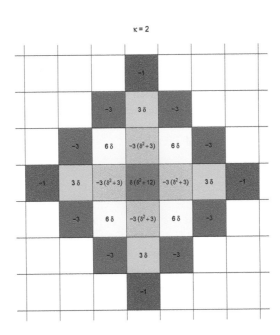

FIGURE 3.11
The two plots show the nighbouring cells to a single grid location (red cell) whose entries in the corresponding conditional autoregressive process are different from zero. Each colour identifies entries with the same reported value. Blank cells correspond to an entry value of zero. The upper configuration is

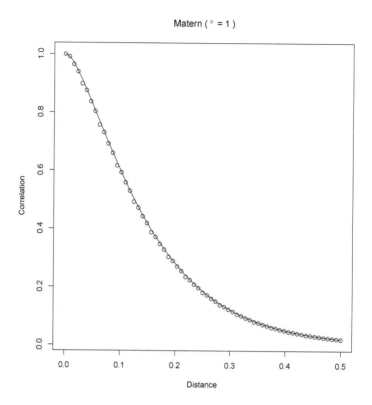

FIGURE 3.12
The Matérn correlation function with scale parameter $\phi = 0.1$ and smoothness parameter $\kappa = 1$(solid lide). The points correspond to the correlation of a CAR process defined over a regular grid with spacing $1/64$ and precision matrix given by the upper panel of Figure 3.11.

is then given by

$$S(x) \approx \tilde{S}(x) = \sum_{k=1}^{n} \psi_k(x) W_k, x \in A, \tag{3.27}$$

where the $\psi_k(\cdot)$ are piecewise linear basis functions defined by the triangulation and $W = (W_1, \ldots, W_k)$ is a zero-mean multivariate Gaussian variate with covariance matrix, Q^{-1} say, chosen to give the required approximation. For example, for $\kappa = 1$, we obtain

$$Q = \left(\frac{\phi^{2\kappa} \Gamma(\kappa)}{4\pi\sigma^2 \Gamma(\kappa+2)} \right)^2 (\phi^{-4}C + 2\phi^{-2}G_1 + G_2)$$

where C, G_1 and G_2 are sparse matrices whose entries are non-zero only for pairs of vertices which share the same triangles.

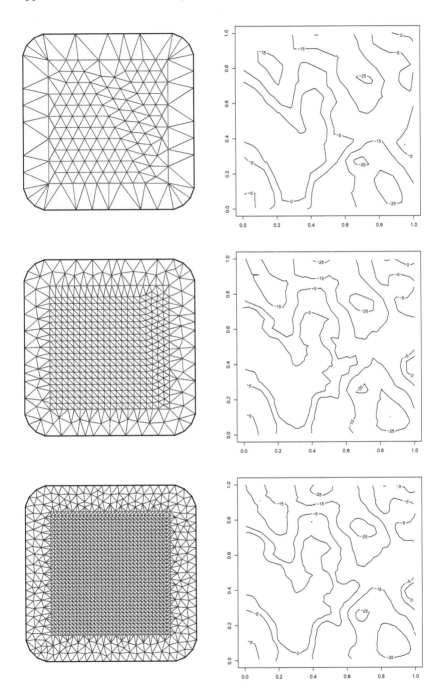

FIGURE 3.13
Left panel: triangulations within the unit square under different constraints.
Right panel: resulting piece-wise linear approximation of a continuous surface.
The solid blue line is the domain polygon.

The accuracy of the approximation in (3.27) is largely dependent on the triangulation of the region of interest A. As the triangulationbecomes fine, the approximation becomes more accurate but also more computationally demanding. Figure 3.13 shows three examples where the region of interest, A, is the unit square. This is approximated in each of the three examples by the solid blue line. It can be seen that additional vertices are also placed outside the unit square in order to avoid so-called edge effects. More specifically, according to Lindgren & Rue (2015), "the boundary effect is negligible at a distance $\rho = \phi(8\kappa)^{1/2}$ from the boundary, and the variance is inflated near the boundary by a factor 2 along straight boundaries, and by a factor 4 near right-angled corners." Hence, it is advisable to extend the domain of the triangulation by a distance at least ρ and to avoid sharp corners so as to mitigate the boundary effect. The function `inla.mesh.2d` from the INLA (Lindgren & Rue, 2015) package in R has been designed to do exactly this. The `PrevMap` package also implements the same GMRF approximation to Matérn processes for both likelihood-based and Bayesian inference, and accepts meshes generated by `inla.mesh.2d` as input.

Note that the contour images in the middle and lower right hand panels of Figure 3.13 are almost identical, indicating that the finest of the three meshes shown in the left hand panels is sufficently fine for all practical purposes. Sections 7.3 Section 7.4 will include applications of this approach in the context of geostatistical inference under preferential sampling.

4

The linear geostatistical model

CONTENTS

In this chapter, we focus our attention on the analysis of measurement data using linear geostatistical models. More specifically we provide answers to the following questions. How can we formulate an appropriate geostatistical model for the data? What methods of inference should we use? How can we predict the outcome of interest at unsampled locations? How can we check that the chosen spatial covariance function is compatible with the data?

We refer readers with a stronger interest in computational issues to Chapters 6 and 7 of Diggle & Ribeiro (2007).

4.1 Model formulation

Formulating a model for a geostatistical data-set is equivalent to specifying a joint probability distribution for the random variable $Y = (Y_1, ..., Y_n)$, indexed by a finite number of unknown parameters, say θ. To pursue this objective, we distinguish between a model for the spatial process S, which represents the state of nature, and a model for the probability distribution of the data Y conditionally on S. Using the same shorthand notation as in Chapter 3, we write this as

$$[Y, S; \theta] = [S; \theta][Y|S; \theta]. \tag{4.1}$$

In (4.1), S is the focus of scientific interest and the data Y, observed at a finite set of locations $\{x_1, \ldots, x_n\}$, represent the information available to us for inference about S, either its properties (parameter estimation) or its realisation (prediction).

The class of linear geostatistical models is then obtained by assuming that, conditional on S, the Y_i are mutually independent with Gaussian conditional distributions. In a slight abuse of notation, we write this as

$$[Y_i|S; \theta] \sim N\left(d(x_i)^\top \beta + S(x_i), \tau^2\right), \text{ for } i = 1, \ldots, n, \tag{4.2}$$

where now $S(x)$ has mean zero and $d(x)$ is vector of explanatory variables at location x with associated regression coefficients β.

The assumption that $S(x)$ is stationary and isotropic implies that

$$\text{Cov}\{S(x), S(x')\} = \sigma^2 \rho(u; \phi),$$

where σ^2 is the variance of the process S, u is the Euclidean distance between locations x and x' and $\rho(u; \phi)$ is a correlation function with parameters ϕ. We usually specify $\rho(u; \phi)$ to be a member of the Matérn family, for the reasons given in Section 3.2.2.

Using the linear properties of Gaussian distributions, we can re-write the geostatistical linear model (4.2) in the following more compact form,

$$Y_i = d(x_i)^\top \beta + S(x_i) + Z_i, \text{ for } i = 1, \ldots, n, \tag{4.3}$$

where Z_i are i.i.d. Gaussian variables with mean zero and variance τ^2. We write the complete set of parameters of the model as $\theta = (\beta, \sigma^2, \phi, \tau^2)$.

In (4.3), the residual component $S(x_i) + Z_i$ represents the variation in Y_i that is not explained by the explanatory variables $d(x_i)$. This residual variation is divided into two components, one spatially correlated the other not. Ideally, $d(x_i)$ would include all the variables that contribute to the spatial structure of Y_i, in which case the $S(x)$ term would be redundant, but in our experience this is rarely the case. When the effect of an explanatory variable is of direct scientific interest it should always be considered for inclusion in the model. However, explanatory variables whose effects are not of direct interest can also be useful; by explaining part of the variation that would otherwise be attributed to $S(x_i)$, their inclusion reduces the uncertainty in predictions based on the fitted model.

The second residual component, Z_i, often referred to as the *nugget effect*, has an ambiguous interpretation, either as spatial variation on a scale smaller than the minimum observed distance or as intrinsic random variation due to measurement error. The first of these two interpretations can be better understood if we define Z to be a spatial process $Z(x)$ with covariance function

$$\text{Cov}\{Z(x), Z(x')\} = \begin{cases} \tau^2 & : \|x - x'\| = 0 \\ \tau^2 \alpha \delta(u) & : \|x - x'\| > 0 \end{cases},$$

where $0 < \alpha < 1$, $\delta(u)$ is a correlation function that takes the value zero for all for u greater than a distance u_0. If the minimum distance between any two data-locations x_i and x_j is at least u_0, the correlation between any two measured values $Z(x_i)$ and $Z(x_j)$ is zero, from which it follows that this model would yield the same likelihood function as when Z is uncorrelated Gaussian noise with no spatial structure. Consequently, neither the parameter α nor the correlation function $\delta(u)$ can be identified, unless the data are augmented to include pairs of sampling locations x_i separated by distances less than u_0 and, ideally, multiple, distinct measurements at the same location.

It follows from the above example that in general, disentangling the two interpretations of the nugget effect is empirically impossible. However, the scientific context of a particular study might help to resolve the ambiguity. For example, if we know that the device used to measure Y_i has a very high precision, then most of the variation in Z_i can be ascribed to small-scale spatial variation. This issue has implications for the design of geostatistical studies, which we shall take up in Chapter 6.

Another, and more basic, aspect of geostatistical design concerns the choice (if choice there be) of the sampling locations x_i. These need not be sampled at random over the region of interest, and we shall explain in Chapter 6 why a completely random sample is not the best choice, even if feasible. The minimal reqirement for the validity of the methods of inference used in the bulk of this book is that the sampled locations should be represetative of the region of interest. Expressed more formally, there should be no stochastic dependence between the sampling process that generates the locations and the scientific process that generates the phenomenon being studied. We shall explore this issue in more detail in Chapter 7.

4.2 Inference

We now describe likelihood-based and Bayesian methods of inference for the linear model. Readers who are willing to take the technical details on trust can skip this section.

4.2.1 Likelihood-based inference

Equation (4.1) is an example of a wide class of so-called *hierarchical* statistical models, in which the distribution of an observable set of random variables,

$Y^\top = (Y_1, \ldots, Y_n)$, is specified conditionally on an unobservable, or *latent* process S. It follows that the likelihood function of the model's parameters θ, given an observed data-set $y^\top = (y_1, \ldots, y_n)$, is obtained by integrating out the process S, hence

$$L(\theta) = \int [S; \theta][y|S; \theta] \, dS. \tag{4.4}$$

For the linear geostatistical model, the integral on the right-hand side of (4.4) is analytically tractable, leading to the following explicit expression for the likelihood,

$$L(\theta) = (2\pi)^{-n/2} |\Sigma|^{-1/2} \exp\left\{ -\frac{1}{2}(y - D\beta)^\top \Sigma^{-1}(y - D\beta) \right\}, \tag{4.5}$$

where D is an n by p matrix of explanatory variables and the n by n variance matrix Σ has $(i, j)^{\text{th}}$ element

$$\Sigma_{ij} = \begin{cases} \sigma^2 \rho(u_{ij}; \phi) & \text{if } i \neq j \\ \sigma^2 + \tau^2 & \text{if } i = j \end{cases}.$$

A useful re-parametrization of this model, for reasons explained in Section 4.2.1.1, is obtained by rewriting $\Sigma = \sigma^2 V = \sigma^2 (R + \nu^2 I)$, where $\nu^2 = \tau^2/\sigma^2$. The log-likelihood for $\theta^\top = (\beta^\top, \sigma^2, \phi^\top, \nu^2)$ is then given by

$$\begin{aligned} l(\theta) &= \log L(\theta) \\ &= -\frac{1}{2}\left[n \log(2\pi\sigma^2) + \log|V| + \frac{(y - D\beta)^\top V^{-1}(y - D\beta)}{\sigma^2} \right]. \end{aligned} \tag{4.6}$$

4.2.1.1   Maximum likelihood estimation

To estimate θ, an obvious strategy is to pass the function $l(\theta)$ to a general-purpose numerical optimization algorithm and so obtain the the maximum likelihood estimate, $\hat{\theta}$. The computational cost of the optimization can be reduced by using a gradient-based optimization algorithm. This is particularly useful in the presence of flat likelihoods, which require numerically accurate differentiation to avoid false convergence. For the linear geostatistical model, the required gradient information is given by the following analytical expressions:

$$\frac{\partial l}{\partial \beta} = -\frac{D^\top V^{-1}(y - D\beta)}{\sigma^2}$$

$$\frac{\partial l}{\partial \sigma^2} = -\frac{1}{2}\left[\frac{1}{\sigma^2} - \frac{(y - D\beta)^\top V^{-1}(y - D\beta)}{\sigma^2} \right]$$

$$\frac{\partial l}{\partial \theta_h^*} = -\frac{1}{2}\left[\text{tr}(V^{-1}V_h) - \frac{(y - D\beta)^\top V^{-1}V_h V^{-1}(y - D\beta)}{\sigma^2} \right]$$

where
$$V_h = \frac{\partial V}{\partial \theta_h^*}$$

and θ_h^* is the h-th element of the augmented vector of parameters $\theta^* = (\phi, \nu^2)$.

Another way to reduce the computational effort involved in the maximisation of $\ell(\theta)$ is to reduce the dimensionality of the optimization problem as follows. Given any fixed value of θ^*, the maximum likelihood estimates of the remaining parameters β and σ^2 can be written down explicitly as

$$\hat{\beta}(\theta^*) = (D^\top V^{-1} D)^{-1} D^\top V^{-1} y \tag{4.7}$$

and

$$\hat{\sigma}^2(\theta^*) = \frac{1}{n}(y - D\beta(\theta^*))^\top V^{-1}(y - D\beta(\theta^*)). \tag{4.8}$$

By plugging these into (4.6), we obtain the function

$$l_p(\theta^*) = -\frac{1}{2}\left[n \log\left\{2\pi\hat{\sigma}^2(\theta^*)\right\} + \log|V| + n\right], \tag{4.9}$$

called the *profile log-likelihood* for θ^*. Numerical optimization algorithms can be applied to $l_p(\theta^*)$ to find the maximum likelihood estimate $\hat{\theta}^*$. The maximum likelihood estimates of β and σ^2 then follow by back-substitution of $\hat{\theta}^*$ into (4.7) and (4.8). Note in particular that in this approach, the dimensionality of the numerical optimization problem does not depend on the number of explanatory variables in the model.

4.2.2   Bayesian inference

In Bayesian inference, the vector of model parameters, θ, is treated as an unobserved random variable and must be assigned a probability distribution, called the *prior distribution* and written as $[\theta]$. The prior distribution represents the researcher's uncertainty about θ before data-collection. Inference about θ is carried out through the distribution of θ given the data, y, called the *posterior distribution* of θ. In the Bayesian formulation of the inference problem, the hierarchical specification (4.1) therefore has to be extended to

$$[Y, S, \theta] = [\theta][S|\theta][Y|S, \theta]. \tag{4.10}$$

Integrating (4.10) with respect to S gives

$$[Y, \theta] = [\theta]\int [S|\theta][Y|S, \theta]dS = [\theta][Y|\theta], \tag{4.11}$$

and an application of Bayes' Theorem to (4.11) gives the posterior distribution of θ as

$$[\theta|Y] = \frac{[\theta][Y|\theta]}{[Y]}, \tag{4.12}$$

where

$$[Y] = \int [\theta][Y|\theta]d\theta.$$

Note that $[Y|\theta]$ is algebraically identical to the likelihood function, $L(\theta)$. Our deliberate change of notation is intended to emphasise the difference in the status of θ between the likelihood-based and Bayesian approaches to inference: an unknown constant in the former, an unobserved random variable in the latter.

Choosing a prior distribution for θ is not straightforward. The right hand side of (4.12) shows that the posterior distribution, and hence any inference about θ, balances information from two sources: the data, expressed through the likelihood, and the chosen prior for θ. In general, the balance between these two sources of information shifts in favour of the likelihood as the size of the data-set increases, and in favour of the prior as this is increasingly concentrated around a particular value. Ideally, the choice of prior should be informed by subject-matter knowledge, but translating this into a unique prior is difficult. For example, consider a scenario in which a researcher is "confident" that the value of a non-negative parameter θ lies between 1 and 10 and that we take "confident" to mean "with 95% probability." Figure 4.1 shows the density functions of three candidate priors for θ, specifically a log-Gaussian, a Gamma and a Uniform distribution, all three of which satisfy the stated prior knowledge. However, within the range 1 to 10, they express different prior beliefs about where θ lies within this range. For example, the log-Gaussian distribution assigns a probability of about 19% that θ lies between 5 and 10, whereas the Gamma and Uniform distributions assign probabilities of about 30% and 52%, respectively, to the same interval. In this hypothetical example, expert opinion might nevertheless help us to select the most appropriate shape for the prior distribution.

In the absence of expert prior knowledge, or when prior knowledge is vague, a pragmatic approach is to use a *diffuse* prior (see Section 3.4), with the aim of ensuring that inference is largely driven by the likelihood, $[Y|\theta]$.

When the prior $[\theta]$ and the posterior $[\theta|y]$ belong to the same family of probability distributions, we call $[\theta]$ a *conjugate prior*. Conjugate priors are not always available, but when they are they greatly simply the computations associated with Bayesian inference, and for this reason can be considered as another pragmatic choice.

Within the linear geostatistical model, conjugate priors are available for (β, σ^2) holding the covariance parameters θ^* fixed. Section 7.2.1 of Diggle & Ribeiro (2007), gives details, and recommends combining these with conjugate priors with discrete priors for the elements of θ^*. This has the computational advantage that samples from $[\theta|y]$ can be drawn using exact sampling methods. More generally, when conjugate priors are not available or are considered inappropriate as expressions of prior belief, the implementation of Bayesian methods involves Markov chain Monte Carlo (MCMC) algorithms or analytical numerical approximation.

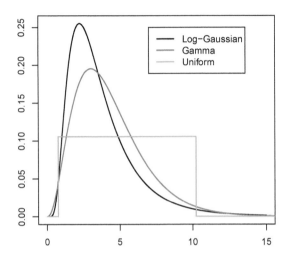

FIGURE 4.1
Log-Gaussian, Gamma and Uniform density functions, $f(\theta)$. In all three cases, $\int_1^{10} f(\theta)d\theta = 0.95$.

4.2.3 Trans-Gaussian models

The fit of the linear geostatistical model can often be improved by applying a transformation to the response variable, Y. Common choices are a power transformation or a log-transformation. The log-transformation is especially common for applications in the life sciences because many biological processes operate multiplicatively rather than additively. Box & Cox (1964) developed a formal inferential methodology based on the family of transformations given by

$$Y_i^* = \begin{cases} (Y_i^\lambda - 1)/\lambda & : \lambda > 0 \\ \log Y_i & : \lambda = 0 \end{cases}$$

This Box-Cox family is essentially the family of power transformations, but defined so that it includes the widely used log-transformation as the limiting case when λ approaches zero.

Cressie (1991) introduced the term *trans-Gaussian model* to refer to the linear geostatistical model with a Box-Cox transformed response variable. In this model, the log-likelihood for $\theta = (\beta, \sigma^2, \phi, \tau^2, \lambda)$ is

$$
l(\theta) = (\lambda - 1) \sum_{i=1}^{n} \log y_i +
$$

$$
- \frac{1}{2} \left[n \log(2\pi\sigma^2) + \log |V| + \frac{(y^* - D\beta)^\top V^{-1}(y^* - D\beta)}{\sigma^2} \right].
$$

The first term on the right-hand side of (4.13) arises from the Jacobian of the transformation from Y to Y^*. Maximization of the likelihood can then be carried out using either of the approaches described in the previous two sections. Direct maximization using gradient information requires the additional partial derivative

$$
\frac{\partial l}{\partial \lambda} = \sum_{i=1}^{n} \log y_i - \mathbf{1}^\top \frac{V^{-1}}{\sigma^2} \left[(y^* - D\beta) \odot \frac{\lambda y^\lambda \odot \log y - y^\lambda + 1}{\lambda^2} \right],
$$

where $\odot$ is the element-wise product between matrices and $\mathbf{1}$ is a vector of length n with all entries equal to 1.

Since λ is usually a nuisance parameter, an alternative and more pragmatic approach is to restrict λ to a discrete set of values, for example $\lambda = -1$ (reciprocal), $\lambda = 0$ (logarithm), $\lambda = 1/2$ (square-root) and $\lambda = 1$ (no transformation). In the authors' experience, restricting the transformation in this way also makes it easier to explain the model to practitioners.

4.3   Model validation

Validation of a statistical model consists of applying diagnostic procedures to check whether the data are compatible with the assumptions incorporated into the model. Most diagnostic procedures involve analysing the residuals from the fitted model. For example, in Section 2.3.1 we developed a Monte Carlo procedure based on the variogram to test for residual spatial correlation in the standard linear model. In this section, we extend the Monte Carlo procedure to provide a check on the validity of the fitted functional form for the spatial correlation. To do this, we first calculate $\hat{V}_0(u)$, the variogram of the residuals from the ordinary least squares fit of the regression component of the model, $d(x)^\top \beta$, to the original data. Now, fix the model parameters θ at their maximum likelihood estimate or, if Bayesian inference is used, the mean of their posterior, $[\theta|y]$, and proceed as follows.

1. Simulate the Gaussian process $S(x_i)$ at each of the observed locations $x_i : i = 1, \ldots, n$.

2. Simulate $Y_i : i = 1, \ldots, n$ as independent realisations of Gaussian random variables with means $d(x_i)^\top \beta + S(x_i)$ and variance τ^2.

3. Compute the residuals $\hat{Z}_i := 1, ..., n$ from the ordinary least squares regression of the simulated Y_i on $d(x_i)$.

4. Compute the variogram, $\hat{V}_1(u)$, based on $\hat{Z}_i$.

5. Repeat steps 1 to 4 a large number of times, say B, to give variograms $\hat{V}_b(u) : b = 1, ..., B$.

Informally, if $\hat{V}_0(u)$ falls within the spread of the simulated $\hat{V}_b(u) : b = 1, ..., B$, this indicates compatibility between model and data, and conversely. A pointwise tolerance band can be calculated by ordering the $\hat{V}_b(u) : b = 1, ..., B$ at each distance u and discarding the most extreme values. For example, to obtain 95% tolerance, we discard the lowest 2.5% and highest 2.5% of the $V_b(u)$ at each distance u.

For a more formal goodness-of-fit test, we need to define a single test statistic. One possibility is

$$T = \sum_{k=1}^{K} |N(u_k)| \left[\hat{V}(u_k) - V(u_k; \phi) \right]^2, \tag{4.13}$$

where $V(\cdot; \phi)$ is the theoretical variogram given by (3.10). Because the sampling variance of the empirical variogram generally increases with u, it is advisable to restrict the summation in (4.13) to relatively small spatial distances u_k. One rule that we have found reasonably effective, and which we illustrate below, is to include only distances up to half the so-called *practical range* of the fitted spatial correlation, conventionally defined as the value of u such that $\rho(u) = 0.05$.

To illustrate the performance of this diagnostic check, we now consider three different scenarios using simulated data on a 32 by 32 regular grid over the unit square. In each case we fit and carry out a goodness-of-fit check for both the correctly specified model and a miss-specified model. The results of the graphical diagnostic are shown in Figure 4.2, whilst Table 4.1 gives the p-values of the formal test based on (4.13).

4.3.1 Scenario 1: omission of the nugget effect

The true model is $Y_i = \mu + S(x_i) + Z_i$, with $\mu = 1$, $\sigma^2 = 1$, $\tau^2 = 1$ and $\rho(u; \phi) = \exp(-u/\phi)$ with $\phi = 0.1$. The miss-specified model is $Y_i = \mu + S(x_i)$, omits the spatially unstructured variation Z_i. Fitting the correctly specified model results in the empirical variogram being contained within the 95% tolerance band (Figure 4.2, upper right panel). For the miss-specified model the empirical variogram in Figure 4.2 lies outside the 95% tolerance band for distances less than 0.2 (Figure 4.2, upper left panel). The formal test based on (4.13) gives a p-value less than 0.001. The miss-specified model estimates the variance of Y_i correctly, but without the option to decompose the variance into spatially structured and unstructured components, it severely underestimates the scale parameter of the spatial correlation, ϕ.

4.3.2 Scenario 2: miss-specification of the smoothness parameter

The true model is $Y_i = \mu + S(x_i) + Z_i$ with $\mu = 1$, $\sigma^2 = 1$, $\tau^2 = 1$, and Matérn correlation function with scale parameter $\phi = 0.01$ and smoothness parameter $\kappa = 5$, whilst the miss-specified model has $\kappa = 0.5$. Recall that in this case the spatial process $S(x)$ is five times differentiable when $\kappa = 5$, mean-square continuous but not differentiable when $\kappa = 0.5$. The correctly specified model again shows a good fit to the data, as expected, whereas for the miss-specified model the graphical diagnostic indicates a poor fit at small distances, $u < 0.1$. The p-value from 4.13 is 0.026.

In our experience, the parameter κ is generally very difficult to estimate, especially when data are sparsely sampled in space. In this synthetic example, the arrangement of more than 1,000 sampling locations over a finely spaced grid leads to data that are able to discriminate between the correct and incorrect values of κ. Even so, the empirical evidence against the miss-specified model is weaker than we found in Scenario 1.

4.3.3 Scenario 3: non-Gaussian data

The true model is $Y_i = \exp\{\mu + S(x_i) + Z_i\}$ with $\mu = 1$, $\sigma^2 = 2$, $\tau^2 = 1$ and $\rho(u; \phi) = \exp\{-u/0.2\}$. The miss-specified model is $Y_i = \mu + S(x_i) + Z_i$, with the same parameterisation. Both the graphical procedure and the formal test give overwhelming evidence of the incompatibility between the miss-specified model and the data.

In this case, the miss-specification can be corrected by a marginal transformation of the data. This serves as a reminder that classical, non-spatial methods of exploratory analysis remain usesful in geostatistical modelling, to explore issues such as the need to transform a response variable, and to check for non-linearity of regression relationships before fitting a spatial model.

Scenario	Misspecified model	True model
1	< 0.001	0.370
2	0.026	0.641
3	< 0.001	0.860

TABLE 4.1
P-values based on the test statistic (4.13) for the misspecified and true model under each of the three scenarios in Section 4.3.1 to 4.3.3.

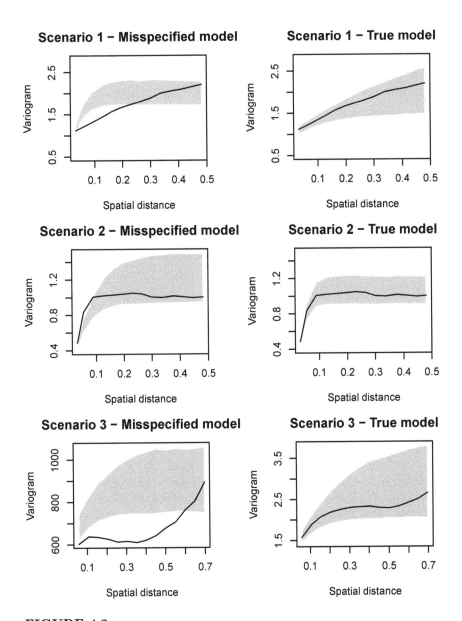

FIGURE 4.2
Results from the graphical test based on the algorithm described in Section
4.3 for each of the models of Section 4.3.1 to 4.3.3. In each panel, the solid line
is the empirical variogram of the residuals from a standard linear regression
model. The shaded area is a 95% pointwise tolerance band generated under
the fitted geostatistical model.

Daniel G. Krige

The term *kriging* acknowledges Daniel G. Krige (1919-2013), whose pioneering work with the South African mining industry in the use of statistical methods to assess ore reserves was later formalised and extensively developed by Georges Matheron and colleagues at the Paris School of Mines, Fontainebleau, France.

4.4 Spatial prediction

We now show how model-based geostatistical prediction relates to the classical geostatistical approach known as *kriging* (see text-box on Daniel G. Krige).

We assume for the time being that the values of all parameters are known. This is not a realistic assumption but it allows us to describe the main ideas initially without unnecessary technical complications.

In Section 3.5, we showed how inference for a predictive target $T = t(Y)$ is carried out through its predictive distribution, $[T|Y]$, i.e. the conditional distribution of what we want to predict (T) given what we know (Y). The expectation of $[T|Y]$, denoted by $E[T|Y]$, minimises the mean square prediction error; see Theorem 3.5.1. In classical geostatistics, with $T = S(x)$ at an arbitrary location x, $E[T|Y]$ is called the *simple kriging* predictor.

We defined the linear Gaussian model in Chapter 3, equation (3.3). We reproduce the equation here for convenience:

$$Y_i = \mu(x_i) + S(x_i) + Z_i : i = 1, ..., n,$$

where $\mu(x) = \alpha + \beta d(x)$, $d(x)$ is a vector of explanatory variables associated with location x, $S(x)$ is a zero-mean Gaussian process and the Z_i are independent and Normally distributed with mean zero and variance τ^2.

Now suppose that our objective is to predict jointly the values of $T(x) = \mu(x) + S(x)$ at any specified set of locations within the region of interest, hence our predictive target is the random vector $T^* = (T(x_1^*), ..., T(x_q^*))^\top$, where $X^* = \{x_1^*, ..., x_q^*\}$ denotes the set of locations for which predictions are required. A convenient shorthand notation is to write $T^* = \mu^* + S^*$ where

$$\mu^* = (d(x_1^*)^\top \beta, \ldots, d(x_q^*)^\top \beta)^\top$$

and

$$S^* = (S(x_1^*), \ldots, S(x_q^*))^\top.$$

Note that in order to predict T^*, the vector of explanatory variables $d(x)$ must be available for all $x \in X^*$. The key result for spatial prediction within the linear Gaussian model is that the conditional distribution of T^* given Y is multivariate Normal. Specifically,

$$E[T^*|y] = \mu^* + \sigma^2 C^\top \Sigma^{-1}(y - \mu), \tag{4.14}$$

where C is the n by q matrix with i^{th} column

$$c_i = (\rho(\|x_i^* - x_1\|; \phi), \ldots, \rho(\|x_i^* - x_n\|; \phi))^\top,$$

and

$$\text{Cov}[T^*|y] = \Sigma^* - \sigma^4 C^\top \Sigma^{-1} C \tag{4.15}$$

where the matrix Σ^* has (i, j)-th element $\Sigma_{ij}^* = \sigma^2 \rho(\|x_i^* - x_j^*\|; \phi)$.

The special case of these results for a single prediction location x gives a useful insight into the way the data contribute to the prediction. Firstly, the point predictor of $T(x)$ is

$$E[T(x)|Y = y] = \mu(x) + \sigma^2 c(x)^\top \Sigma^{-1}(y - \mu(x)) \tag{4.16}$$

where

$$c(x) = (\rho(\|x - x_1\|; \phi), \ldots, \rho(\|x - x_n\|; \phi))^\top. \tag{4.17}$$

Secondly, the prediction variance is

$$v(x) = \text{Var}[T(x)|Y = y] = \sigma^2(1 - \sigma^2 c(x)^\top \Sigma^{-1} c(x)). \tag{4.18}$$

Equation (4.16) can be re-expressed as

$$\hat{T}(x) = \mu(x) + \sum_{i=1}^{n} w_i(x)(Y_i - \mu(x_i)),$$

where the weights $w_i(x)$ depend on three things: x; $\mu(x)$; and the covariance structure of the Gaussian process $S(x)$. The first term on the right-hand-side of (4.16), $\mu(x)$, is the best prediction we could make if we only observed the explanatory variables $d(x)$ or if, improbably, all of the Y_i were equal to their expectations, $\mu(x_i)$. The second term on the right-hand-side of (4.16) expresses the deviation of the predictor, $\hat{T}(x)$ from $\mu(x)$ as a weighted average of the deviations of the Y_i from *their* expectations, $\mu(x_i)$, in which the weights, $w_i(x)$, depend on the values of the correlation between $S(x)$ and $S(x_i)$. The exact form of this dependence is subtle, involving the configuration of the data-locations x_i in ways that are not always obvious. Typically, and as intuition might suggest, the $w_i(x)$ tend to be larger for locations x_i closer to the target prediction location x. It follows that if the observed values of the Y_i at locations close to x are larger than their expectations, the predicted value $\hat{T}(x)$ will typically be larger than *its* expectation. Another consequence of (4.17) is that if the prediction location x is sufficiently remote from all of the data-locations x_i so that all n correlations $\rho(\|x - x_i\|)$ are zero, then all of the weights $w_i(x)$ are zero. Again, this is intuitively reasonable: if all of the data are uncorrelated with $T(x)$, the best prediction we can make for $T(x)$ is its expected value.

Now, consider the prediction variance, as given by (4.18). Notice that this does not depend on the values of the Y_i, but only on the configuration of the data-locations and the covariance structure of the data. Also, the prediction variance is never bigger than σ^2, the variance of $T(x)$.

The prediction variance $v(x)$, or more helpfully the prediction standard deviation $\sqrt{v(x)}$, measures the precision of the point predictor $\hat{T}(x)$. A more complete measure of precision is the conditional distribution of $T(x)$ given Y, called the *predictive distribution* of $\hat{T}(x)$. Because $\hat{T}(x)$ is a linear combination of the Y_i and Y is multivariate Gaussian, the predictive distribution is also Gaussian, with mean $\hat{T}(x)$ and variance $v(x)$. From this we can make probability statements about the unknown $T(x)$. For example, in air pollution studies a policy-relevant statement would be the probability that air pollution at location x falls below a regulatory limit.

These results can also be used to produce a predictive map of $T(x)$ throughout the region of interest, A. We simply use equations (4.16) and (4.18) repeatedly for locations x that form a grid to cover A at the required spatial resolution for the map. However, it is important to note that this only allows us to make marginal probability statements about $T(x)$, because predictions at different locations are not independent. To make a probability statement about more than one location at a time, we need to use the more general formulae (4.14) and (4.15).

A final observation on equations (4.16) and (4.18) concerns their behaviour when the prediction location x coincides with one of the data-locations x_i. In this case, if $\tau^2 = 0$, then necessarily $\hat{T}(x_i) = y_i$ with prediction variance zero, because y_i is identically equal to $\mu(x_i) + S(x_i)$. In other words, the kriging predictor interpolates the data, not unreasonably so if the data are observed without error. If, on the other hand, $\tau^2 > 0$, then the observed fluctuations of the y_i relative to their expectations $\mu(x_i)$ will be smoothed out, to an extent that depends on the covariance structure of the model, but primarily on the noise-to-signal ratio, τ^2/σ^2.

In general, the predictive target T can be any property of the realised spatial surface over the region of interest, A. We write this as $T = t(\mu*, S*)$ where μ^* and S^* denote the sets of values of $\mu(x)$ and $S(x)$, respectively, as x runs through A, which in practice we approximate by a fine grid of locations to cover A. The meaning of "fine" is context-dependent and may need to be chosen by trial-and-error. Using the grid to approximate the spatially continuous surface $\mu(x) + S(x)$ is equivalent to assuming that $\mu(x) + S(x)$ is approximately constant within grid-cells, and the grid therefore needs to be fine enough that the effect of this approximation is of a smaller order of magnitude than the inherent imprecision of the predictor $\hat{T}$.

For linear functions $t(\cdot)$, for example spatial averages of $\mu(x) + S(x)$ over sub-regions of A, the predictive distribution of T is Normal, with mean and variance that can easily be calculated using (4.14) and (4.15). For non-linear functions $t(\cdot)$, the point predictor $E[T|y]$ cannot be calculated explicitly. Instead, we use the following Monte Carlo method.

Again using locations $x_1^*, \ldots, x_q^*$ to form a fine grid over A, draw B independent samples, $S_b^* = (S_b(x_1^*), \ldots, S_b(x_q^*))^\top$, from the multivariate Gaussian distribution of S^* given Y; this has mean vector $\sigma^2 C^\top \Sigma^{-1}(y - \mu)$ and covariance matrix given by (4.15). It follows that $T_b = t(\mu^*, S_b^*)$, for $b = 1, \ldots, B$, are an independent random sample from the predictive distribution of T, and if B is large enough their empirical distribution can be used to make any required probability statement about T. Again, "large enough" is context-dependent but typical values might be $B = 1{,}000$ or $10{,}000$.

If we want a point predictor $\hat{T}$, we approximate the conditional expectation of T given Y by the sample mean of the simulated values, hence

$$\hat{T} = \frac{1}{B} \sum_{b=1}^{B} T_b.$$

Similarly, we approximate the prediction variance by the sample variance

$$\mathrm{Var}[T|y] \approx \frac{1}{B-1} \sum_{b=1}^{B} (T_b - \hat{T})^2.$$

One circumstance in which prediction is very naturally focused on a non-linear function is for a trans-Gaussian model where, typically, scientific interest is in the untransformed phenomenon. For example, if the linear Gaussian model is fitted to log-transformed data, the spatial surface of scientific interest is $T(x) = \exp\{\mu(x) + S(x)\}$. In this case, the predictive distribution of $T(x)$ is log-Normal and analytically tractable. In particular, using m and v to denote the mean and variance of the predictive distribution of $S(x)$, the predictive distribution of $\exp\{S(x)\}$ has mean $m^* = \exp(m + v/2)$ and variance

$$v^* = \{\exp(v) - 1\} \exp(2m + v).$$

So far, we have assumed that all parameters are known. We now consider the typical scenario in which they are unknown. Denote the complete set of model parameters by θ. A pragmatic strategy, called *plug-in prediction*, is to replace θ by a point estimate $\hat{\theta}$, for example the maximum likelihood estimate or, if Bayesian inference is being used, the posterior mean. This ignores the uncertainty in $\hat{\theta}$ but often gives good results because parameter uncertainty is typically dominated by predictive uncertainty. An intuitive explanation for this is that all of the data contribute to estimation of θ whereas only data from locations close to a particular location x contribute materially to prediction of $T(x)$.

To check whether parameter uncertainty needs to be taken into account, and to do so if necessary, one option is to use the fact that in large samples, the maximum likelihood estimator $\hat{\theta}$ has a multivariate Normal sampling distribution. To allow for this uncertainty Giorgi et al. (2018) suggested modifying the Monte Carlo method outlined above by first drawing samples $\tilde{\theta}_b : b = 1, \ldots, B$ from this multivariate Normal distribution and then using $\tilde{\theta}_b$ when drawing S_b^*, rather than using the same plug-in value $\hat{\theta}$ for all of the S_b^*. The multivariate Normal approximation to the sampling distribution of θ is improved by transforming the covariance parameters, and we use this strategy in the examples that follow in Section 4.5.

As shown in Section 3.5, Bayesian statistics provides a formal approach to incorporating parameter uncertainty by integrating out the dependence on θ. The Bayesian predictive distribution for any target T is

$$[T|y] = \int [\theta|y][T|y, \theta]\, d\theta,$$

a weighted average of plug-in predictions with weights given by the posterior distribution of θ. To implement this in practice, we again use the above modification of the Monte Carlo method, but now drawing the values $\tilde{\theta}_b$ from the posterior distribution $[\theta|y]$, rather than from the multivariate Normal sampling distribution of the maximum likelihood estimator $\hat{\theta}$.

4.5   Applications

4.5.1   Heavy metal monitoring in Galicia

In this section we analyse data on lead concentration (in ppm, parts per million dry weight) measured in 132 moss samples of the species *Scleropodium purum* collected from a survey conducted in Galicia, Northern Spain, in 2000. The data can be found in the file galicia.csv on the book's web-page. Here, we compare likelihood-based and Bayesian methods of inference relating to parameter estimation and geostatistical prediction.

Figure 4.3 is a point-map of the data, showing the arrangement of the geographical locations where lead concentration was measured. These form a nearly-regular grid. Histograms of the data before and after applying a log-transformation (Figure 4.4) show that a log-transformation produces a more nearly symmetric empirical distribution of the 132 measured lead concentrations.

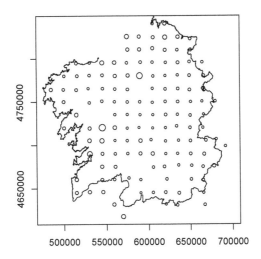

FIGURE 4.3
Point-map of the Galicia lead concentration data. Each circle has center at a sampled spatial location and radius proportional to the measured lead concentration. The continuous line shows the administrative boundary of Galcia.

Lead poisoning

- **Public health issues.** Lead is a naturally occurring metal whose intensive use in industrial processes has resulted in an increase in environmental pollution. This has public health implications because lead has toxic effects on humans, affecting natural brain development in children and causing potential kidney damage in adults.

- **Exposure.** People are exposed to lead through inhalation (e.g. burning of any material containing lead) and ingestion (e.g. lead-contaminated food and water).

- **Global burden.** In 2013, diseases from lead exposure were responsible for an estimated 853,000 deaths globally.

- **Source.** http://www.who.int/mediacentre/factsheets/fs379/en/

The first step in the analysis is to verify that the data are spatially structured. The left-hand panel of Figure 4.5 shows the empirical variogram of

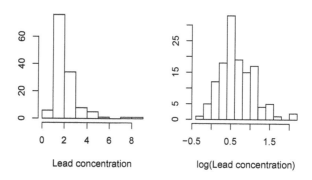

FIGURE 4.4
Histograms of the Galicia lead concentation data(left panel) and log-transformed data (right panel).

log-transformed lead concentrations together with 95% tolerance band derived from 1,000 random permutations. This clearly indicates the existence of spatial correlation, with the empirical variogam falling outside the tolerance envelope at distances up to about 50km. Moreover, the shape of the empirical variogram suggests that an exponential function, $\rho(u) = \exp(-u/\phi)$, is a reasonable model to describe the spatial correlation, with a correlation range of the order of 100m. Finally, the behaviour of the empirical variogram indicates that the nugget term contributes at most a small proportion of the total variance.

The next step is to formulate and fit a Gaussian process model. Let Y_i denote the lead concentration at location x_i. Based on the exploratory variogram analysis, we adopt the following model:

$$\log\{Y_i\} = \mu + S(x_i), \tag{4.19}$$

where $S(x)$ is a stationary and isotropic zero-mean Gaussian process with variance σ^2 and exponential correlation function, $\rho(u) = \exp(-u/\phi)$. Rough initial estimates of the parameters, taken "by eye" from the empirical variogram, are $\tilde{\sigma}^2 = 0.2$, the approximate value of the "sill," and $\phi = 30$, corresponding to a practical range of 90km. Arguably, we could have added a nugget term to the model, to give

$$\log\{Y_i\} = \mu + S(x_i) + Z_i$$

where the Z_i are independent $N(0, \tau^2)$, but we anticipated that the fitted value of τ^2 would be close to zero.

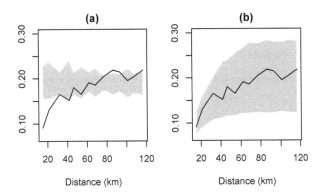

FIGURE 4.5
Galicia lead concentration data. Results from variogram diagnostic check for
the presence of residual spatial correlation (left-hand panel) and for compatibility of the data with the fitted geostatistical model (4.19, right-hand panel).
The solid line is the empirical variogram of the data. The shaded areas are
95% tolerance bands under the hypothesis of spatial independence (left-hand
panel) and under the fitted model (4.19, right-hand panel).

For a Bayesian analysis, we specify the following set of independent priors.

- $\mu \sim N(0, 10^{10})$.

- $\sigma^2 \sim \text{Uniform}(0, 100)$.

- $\phi \sim \text{Uniform}(0, 300)$.

These appear to be non-informative in the sense that they cover implausibly
wide ranges of the respective parameters.

Parameter estimates, standard errors and 95% confidence/credible intervals are reported in Table 4.2, for both likelihood-based and Bayesian inference. The point estimates are similar, but the Bayesian approach shows larger
standard errors and wider interval estimates than those from the likelihood-
based approach. The discrepancies can be explained by the relatively small
sample size, which makes the results from the Bayesian approach sensitive
to the particular choice of priors. However, the results for spatial prediction
carried out under the two paradigms turn out to be very similar. This can be
seen from Figure 4.6 which compares the means (left-hand panel) and standard errors (right-hand panel) over prediction locations in a regular grid at a
spacing of 20km; both scatterplots are very strongly concentrated around the
line of equality.

The two panels of Figure 4.7 show predictive maps generated from the

TABLE 4.2

Galicia lead concentration data. Parameter estimation using likelihood-based and Bayesian methods. The table gives maximum likelihood estimates (MLE), standard errors (Std. Error) and 95% confidence interval for the likelihood-based method, posterior means, posterior standard errors and 95% credible intervals for the Bayesian method.

	Likelihood-based inference		
Parameter	MLE	Std. Error	95% Confidence interval
μ	0.724	0.100	(0.528, 0.921)
$\log(\sigma^2)$	-1.651	0.196	(-2.036, -1.267)
$\log(\phi)$	3.024	0.256	(2.523, 3.525)
	Bayesian inference		
Parameter	Posterior mean	Std. Error	95% Credible interval
μ	0.752	0.214	(0.345, 1.269)
$\log(\sigma^2)$	-1.278	0.387	(-1.827, -0.410)
$\log(\phi)$	3.461	0.443	(2.748, 4.449)

likelihood-based fit. The left hand panel shows predictive mean map of un-transformed lead-concentrations. That is, the mapped value is the expectation $\exp\{S(x)\}$ with respect to the predictive distribution of $S(x)$. This map picks out a large area of relatively high lead concentrations in the south-west, and several smaller areas in the north. However, compliance with air pollution standards is typically based not on average pollution levels, but on whether or not levels exceed a policy-agreed threshold. In this spirit, the right-hand panel of Figure 4.7 maps the predictive probability that local lead concentration exceeds 4ppm dry weight. Unsurprisingly, this probability is relatively large in areas where the predictive mean is relatively large. This is not necessarily so; in this application the close correspondence is a consequence of the grid-like arrangement of the sample locations, with grid-spacing substantially smaller than the correlation range of the fitted model.

This data-set, as provided to us, did not include any explanatory variables. However, two explanatory variables are automatically available with any geostatistical data-set, namely the spatial coordinates, $x = (x_1, x_2)$ of any location in the study-region. In the geostatistical context, a regression model that takes the form of a polynomial in x_1 and x_2 is called a *trend surface*. For example, *linear* and *quadratic* trend-surfaces are of the form

$$\mu(x_1, x_2) = \beta_0 + \beta_1 x_1 + \beta_2 x_2$$

and

$$\mu(x_1, x_2) = \beta_0 + \beta_1 x_1 + \beta_2 x_2 + \beta_3 x_1^2 + \beta_4 x_2^2 + \beta_5 x_1 x_2,$$

respectively.

We incline towards not using a trend surface model unless, in context, a smooth trend in geographical coordinates has a natural interpretation. This

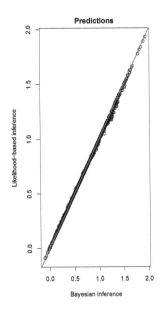
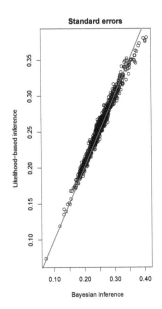

FIGURE 4.6

Spatial prediction for Galicia lead concentration data. Point predictions (left panel) and standard errors (right panel) of the log-transformed lead concentration were computed on a 20 by 20 km regular grid over Galicia. In both plots, the vertical axis corresponds to results from the likelihood-based method, and the horizontal axis to results from the Bayesian method.

might apply if, for example, the study-region covers a large geographical area and the phenomenon of interest is climate-related. Even then, we would advise against fitting anything more complicated than a quadratic surface because high-order polynomial fits tend to include features that bear no relationship to the patterns in the data, especially when data-locations are irregularly spaced. In the case of the Galicia lead pollution data, re-fitting the model (4.19) to the log-transformed lead concentrations with the addition of a linear trend-surface model for $\mu(x)$ makes virtually no difference to the predictive maps shown in Figure 4.7.

4.5.2   Malnutrition in Ghana (continued)

We now continue our analysis of the data on malnutrition in Ghana. In Section 2.1.1 we explored the association of the height-for-age Z-score (HAZ) with age (d_1), maternal education (d_2) and family wealth index (d_3) using a standard linear regression model. The test statistic (2.12) for the presence of

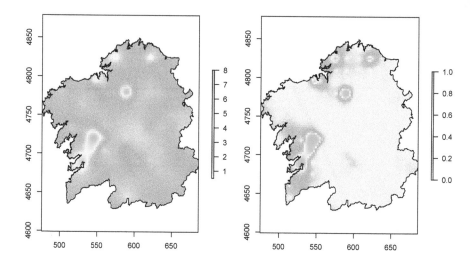

FIGURE 4.7
Spatial prediction of lead concentrations in moss samples, based on the model
(4.19). Predictive mean (left-hand panel) and predictive probability of exceeding 4ppm dry weight (right-hand panel).

residual spatial correlation, yields a p-value less than 0.001. We now develop
a geostatistical model by extending (2.2) to

$$Y_j(x_i) = \beta_0 + \sum_{h=1}^{3} \beta_h b_h(d_{1,ij}) + \beta_4 d_{2,ij} + \beta_5 d_{3,ij} + S(x_i) + U_{ij}, \qquad (4.20)$$

where i denotes cluster and j denotes child within cluster. This model includes
two components of variation: $S(x)$ is a Gaussian process with exponential
correlation function; the U_{ij}, representing child-specific residual variation, are
independent Gaussian variables with mean zero and variance ω^2.

Table 4.3 reports the maximum likelihood estimates of the model parameters. Point estimates and standard errors of the regression coefficients are
similar to those obtained from the non-spatial linear regression in Table 2.1.
At first sight, this is surprising in view of our earlier finding that the residuals
from the non-spatial regression model were spatially correlated. However, it
can be explained by looking at the estimates of the covariance parameters.
These indicate that the child-specific component of residual variation in HAZ
(ω^2), which is accounted for in the non-spatial linear regression analysis, is
bigger than the between-cluster variation (σ^2) by an estimated factor of approximately 20.

The validation of the model based on the algorithm of Section 4.3 does

TABLE 4.3
Maximum likelihood estimates with associated standard errors (Std. Error) and 95% confidence intervals (CI) for the regression coefficients of the linear geostatisical model in (4.20).

Parameter	Estimate	Std. Error	95% CI
β_0	-0.567	0.124	(-0.811, -0.323)
β_1	-0.655	0.130	(-0.910, -0.400)
β_2	-0.196	0.203	(-0.594, 0.202)
β_3	0.978	0.114	(0.755, 1.201)
β_4	0.164	0.060	(0.046, 0.283)
β_5	0.306	0.044	(0.219, 0.392)
$\log(\sigma^2)$	-2.685	0.314	(-3.301, -2.068)
$\log(\phi)$	3.782	0.577	(2.651, 4.914)
$\log(\omega^2)$	0.333	0.635	(-0.912, 1.578)

not show any evidence against the fitted exponential correlation function, with the empirical variogram of the residuals falling within the 95% tolerance bandwidth (Figure 4.8). The test in (4.13) is also consistent with this finding, yielding a p-value of about 0.7. However, this result should be interpreted cautiously, for the following reason. The practical range of the spatial correlation, defined as the distance beyond which the spatial correlation is below 0.05, is estimated to be $\log(20) \times \hat{\phi} = \log(20) \times 80.078 \approx 240$ km, but with very poor precision; the 95% confidence interval for the practical range extends from approximately 57 to 1006 km. The fact that the upper end of this interval is substantially bigger than the east-west linear dimension of Ghana serves as a warning that the spatial covariance parameter estimates will be collectively imprecise. In particular, visual inspection of Figure 4.8 suggests that a statistically acceptable fit could be obtained either by a model whose variogram flattens out at a distance of around 100 to 150km, or by one whose variogram continued to increase beyond the range of distances covered by Figure 4.8; it is to illustrate this ambiguity that we have broken our own rule of thumb about not estimating a variogram at distances comparable to the linear dimensions of the study-region.

Our fitted model for these data includes covariates that are properties of a person at a location, rather than of a location itself. It follows that predictive maps can only be constructed under hypothetical scenarios about the people who live at an unsampled location. For illustration, we suppose that our target for prediction, $T(x)$, is the HAZ of a 1 year old child, having a mother with the lowest level of education and living at location x in a family with the lowest level of wealth index. Hence,

$$T(x) = \beta_0 + \sum_{h=1}^{3} \beta_h b_h(1) + \beta_4 + \beta_5 + S(x), x \in \mathcal{G} \qquad (4.21)$$

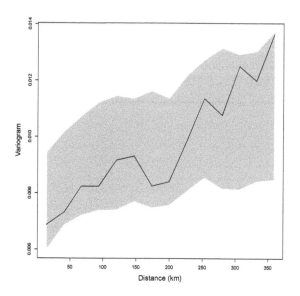

FIGURE 4.8
Results from the diagnostic check on the compatibility of the adopted spatial
correlation function for the malnutrition data. The solid line corresponds to
the empirical variogram of the residuals from the standard linear regression.
The shaded area represente the 95% tolerance bandwidth obtained using the
algorithm described in Section 4.3.

where $\mathcal{G}$ is a 10km by 10 km regular grid covering the whole of Ghana.
Figure 4.9(a) shows the results for the spatial predictions of this target. We
estimate levels of HAZ below -1.6 in the north-east and in a smaller area in
the south-west. Figure 4.9(b) shows the predictive probability that HAZ is
below -2, formally expressed as

$$r(x) = P(T(x) < -2|y), x \in \mathcal{G}.$$

We observe the presence of a small hotspot of stunting risk in the north-east
where $r(x)$ reaches a maximum value of about 70%, while in the rest of the
country $r(x)$ is close to zero.

4.5.2.1 Spatial predictions for the target population

In Figure 4.9, spatial predictions were carried out by fixing age, maternal
education and family wealth at pre-defined values. A more policy-relevant
target population might be the actual population of children in Ghana aged
less than 5 years, who will vary in age, maternal education and wealth index.
How can we generate predictions of HAZ for this target population? As for

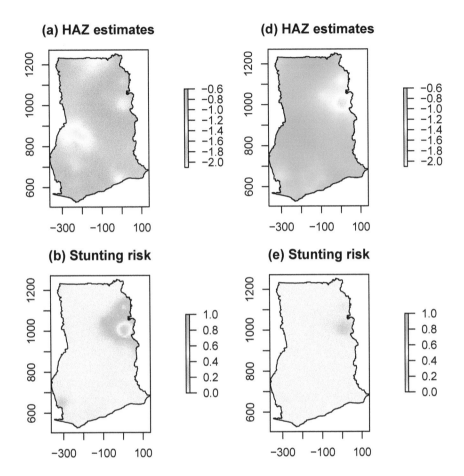

FIGURE 4.9
Results from the spatial prediction of the malnutrition data. Upper panels: predicted surfaces of height-for-age Z-scores (HAZ). Lower panels: maps of the predictive probability that HAZ lies below a threshold of -2. Left panels: age is fixed at 1 year, maternal education and wealth index at their lowest score. Right panels: results for the general population of children in Ghana aged less than 5 years; see Section 4.5.2.1 for more details on how these are obtained. In each of the maps, the results are reported on a 10 by 10 km regular grid covering the whole of Ghana.

any prediction problem, the answer is a probability distribution, but which one?

A complete answer requires either that the values of the explanatory

TABLE 4.4
Empirical joint distribution of maternal education (from 1="Poorly educated"
to 3="Highly educated") and wealth index (from 1="Poor" to 3="Wealthy").

		Wealth index		
		1	2	3
	1	0.701	0.127	0.073
Maternal education	2	0.019	0.018	0.026
	3	0.003	0.007	0.026

variables are known for every member of the target population, which is
infeasible, or that we are willing to make modeling assumptions about the
joint spatial distribution of HAZ, age, maternal education and wealth index
over the target population. Formally, we write $\mathcal{D} = \{D(x) : x \in \mathcal{G}\}$, where
$D(x) = \{D_1(x), D_2(x), D_3(x)\}$, $D_1(x)$ represents age, $D_2(x)$ maternal education and $D_3(x)$ the wealth index. The predictive distribution at any location
x is then obtained by averaging over the distribution of $\mathcal{D}$, to give

$$[T(x)|y] = \int [T(x)|\mathcal{D}, y][\mathcal{D}] \, d\mathcal{D}, x \in \mathcal{G} \qquad (4.22)$$

where $[T(x)|\mathcal{D}, y]$ is a multivariate Gaussian distribution with mean and co-
variance matrix given by (4.14) and (4.15), respectively.

Specifying a model for $\mathcal{D}$ takes us into the realm of multivariate geostatistical modelling, which in our view is best approached by taking account
of features specific to the context of the problem. We shall describe several
applications of this approach in Chapter 10 . Here, we make two rather strong
simplifying assumptions: firstly, that the joint distribution of age, maternal
education and wealth index is the same at all locations; secondly, and more
plausibly, that age is independent of maternal education and wealth index.
Together, these imply that at any location x, the joint distribution of $D_1(x)$,
$D_2(x)$ and $D_3(x)$ can be written as

$$[D_1(x), D_2(x), D_3(x)] = [D_1][D_2, D_3].$$

As our sample of children is large, we estimate the distributions $[D_2, D_3]$ and
$[D_1]$ as their empirical counterparts in our data.

The empirical joint distribution of the categorial variables D_2 and D_3 is
shown in Table 4.4. For the continuous variable D_1, the child's age, we sample
with replacement from its empirical distribution, shown in Figure 4.10.

To compute the integral (4.22) we use a Monte Carlo method by first drawing samples from $[D_1]$ and $[D_2, D_3]$ and then, for each of these, simulating from
the multivariate Gaussian distribution $[T(x)|\mathcal{D}, y]$. Finally, we use the result-
ing set of simulated samples to compute the relevant predictive summaries for
$[T(x)|y]$. The results of this procedure are shown in Figure 4.9(d)-(e).

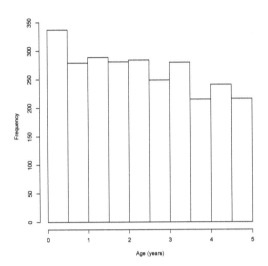

FIGURE 4.10
Histogram of the age distribution in the data on malnutrition.

As expected these are remarkably different from the rather extreme scenario used to generate Figure 4.9(a)-(b). We now see higher predicted levels of HAZ and a lower risk of stunting throughout Ghana, for this estimated general population.

5

Generalised linear geostatistical models

CONTENTS

In Section 2.2 we introduced the class of generalised linear models as an extension of the linear regression model for analysing non-Gaussian data under the assumpton that measurements at different locations are statistically independent of each other. In this chapter, we describe the extension of generalised linear models to the geostatistical setting. The chapter follows the same structure as the previous chapter: model formulation, with a special focus on Binomial and Poisson models; methods of inference, using likelihood-based and Bayesian approaches; model validation; and spatial prediction. Additionally, in the last section, we describe the link between generalised linear geostatistical models and point process models for analysing data from case-control studies. Users whose main interest is in applications may wish to skip the more technical parts of Section 5.2.

5.1   Model formulation

Our starting point, as in Section 2.2, is a hierarchical representation of the joint distribution of a spatial process S and data $Y = (Y_1, ..., Y_n)$, given parameters θ, as

$$[Y, S; \theta] = [S; \theta][Y|S; \theta]. \tag{5.1}$$

In Chapter 4 we assumed that conditional on S, the Y_i were independent, Normally distributed with means $d(x_i)^\top \beta + S(x_i)$ and common variance τ^2. To obtain the class of *generalized linear geostatistical models* (GLGMs), we make three changes, the first two of which are the same as in the classical generalized linear model for independently replicated data, namely: allowing the conditional distribution of Y_i to be non-Normal; allowing $d(x_i)^\top \beta + S(x_i)$ to be any specified function of the conditional expectation of Y_i. The third extension is to incorporate a new set of *independent* random effects U_i, which are Normally distributed with mean zero and variance ν^2. The resulting model has the hierarchical form

$$[Y, S, U; \theta, \nu^2] = [S; \theta][U; \nu^2][Y|S, U; \theta]. \tag{5.2}$$

Recall from Section 2.3 that these independent random effects also featured in a (non-spatial) generalized linear *mixed* model. Hence, in a GLGM observed responses $Y = (Y_1, ..., Y_n)$ are conditionally independent given the realisations of an unobserved Gaussian process $S(x)$ and a set of independent Normally distributed random variables U_i, and the conditional distribution of Y_i has expectation $g(\eta_i)$, where

$$\eta_i \;=\; d(x_i)^\top \beta + S(x_i) + U_i. \tag{5.3}$$

The quantity η_i is called the *linear predictor* and $g(\cdot)$ the *link function* of the model. The U_i term on the right-hand side of (5.3) is analogous to the nugget effect in the linear model. However, a subtle difference is that in the most widely used examples of GLGMs, namely binomial logistic and Poisson log-linear models it is possible to estimate both the covariance structure of the spatial process $S(x)$ and the variance of the independent random variables U_i because in either case the variance of the conditional sampling distribution of Y_i is a specified function of its mean.

The analogous hierarchical representation of a linear Gaussian model including both $S(x)$ and U_i would be of the form (5.2) with $S(x)$ a Gaussian process, U_i independent $N(0, \nu^2)$ and the Y_i conditional on $S(x)$ and U_i Normally distributed with means $\eta_i = d(x_i)^\top \beta + S(x_i) + U_i$ and varince τ^2. The non-hierarchical representation of the same model is

$$Y_i = d(x_i)^\top \beta + S(x_i) + U_i + Z_i \tag{5.4}$$

where the Z_i are mutually independent $N(0, \tau^2)$. However, using the fact that the sum of two zero-mean Normally distributed random variable is also a zero-mean Normally distributed random variable, (5.4) is indistinguishable from the superficially simpler model

$$Y_i = d(x_i)^\top \beta + S(x_i) + Z_i^*$$ (5.5)

where the Z_i^* are mutually independent $N(0, \omega^2)$ and $\omega^2 = \nu^2 + \tau^2$. In other words, the only way we can discriminate between (5.4) and (5.4) is if information other than provided by the data Y_i can be used to assign a value to either ν^2 or τ^2.

In formulating a GLGM for a particular application, an important decision is whether to include $S(x_i)$, U_i or both into the linear predictor η_i. Before addressing this issue, we recall what interpretation we attach to each of the two random effect components in (5.3). The term $S(x_i)$ corresponds to unexplained spatial variation in the Y_i and, if present, induces spatial correlation in the data. In contrast, the term U_i represents unstructured variation in Y_i over and above sampling variation. It is included in the model to capture the effects of unmeasured explanatory variables that either have no spatial structure or are spatially varying but on a scale smaller than the minimum observed distance.

5.1.1 Binomial sampling

In low-to-middle-income-countries (LMICs), where disease registries are non-existent or geographically incomplete, household surveys represent a viable and useful approach to monitoring the burden of infectious diseases; for example, river-blindness, as discussed in Chapter 1. The format of a data-set arising from a study of this kind can be expressed formally as follows. Let x_i denote the location of a sampled community or village in the population of interest. At each location x_i, n_i individuals are selected and tested for presence or absence of the disease under investigation. The number testing positive is Y_i. Additionally, explanatory variables associated with a location x are available, and denoted by $d(x)$. A natural model for the sampling distribution of Y_i is a Binomial distribution with n_i trials and probability of a positive test $p(x_i)$, where

$$\log\left\{\frac{p(x_i)}{1 - p(x_i)}\right\} = d(x_i)^\top \beta + S(x_i) + U_i.$$ (5.6)

A few qualifying remarks are necessary at this point. Firstly, the binomial sampling distribution strictly assumes that the sampled individuals at each location x_i are a completely random sample from the population at risk. In practice, this is not always achievable, in which case the study-design needs to consider how to ensure that the sampled individuals are an "as if random" sample. Secondly, as discussed at the end of Section 4.1 in the context of the linear geostatistical model, the locations x_i need not be sampled at random over the region of interest. Finally, the log-odds, or *logit* link used in (5.6) is a widely used default, but should not be adopted unquestioningly.

Why Bernoulli extra-variation does not exist

Consider a Bernoulli mixed model as in (5.6) and, without loss of generality, set $S(x) = d(x) = 0$ for all x. It then follows that

$$
\begin{aligned}
P[Y_i = 1] &= \int_{-\infty}^{+\infty} [U_i]P[Y_i = 1|U_i]\, dU_i \\
&= \int_{-\infty}^{+\infty} \frac{e^u}{1+e^u} \frac{1}{\sqrt{2\pi\tau^2}} e^{-\frac{u^2}{2\tau^2}}\, du \\
&= \int_{-\infty}^{+\infty} \left(\frac{e^u}{1+e^u} - \frac{1}{2} \right) \frac{1}{\sqrt{2\pi\tau^2}} e^{-\frac{u^2}{2\tau^2}}\, du + \\
&\quad \frac{1}{2} \int_{-\infty}^{+\infty} \frac{1}{\sqrt{2\pi\tau^2}} e^{-\frac{u^2}{2\tau^2}}\, du.
\end{aligned}
$$

The integrand in the first term on the right-hand side of this equation is an odd function whose integral over the real line is therefore zero. The integrand on the second term is a probability density function whose inegral is therefore 1. Also,

$$
\begin{aligned}
P[Y_1 = 1,\ldots,Y_n = 1] &= \int_{\mathbb{R}^n} \prod_{i=1}^{n} [U_i]P[Y_i = 1|U_i]\, dU_1 \ldots dU_n \\
&= \prod_{i=1}^{n} \int_{\mathbb{R}^n} [U_i]P[Y_i = 1|U_i]\, dU_i \\
&= \prod_{i=1}^{n} P[Y_i = 1] = \left(\frac{1}{2} \right)^n.
\end{aligned}
$$

This shows that the Y_i are jointly distributed as mutually independent Bernoulli variables, each with probability $1/2$ of "success", thus the introduction of the U_i has not introduced any extra-variation.

When $n_i = 1$ for all $i = 1,\ldots,n$, the Y_i constitute as set of independent *Bernoulli trials*, i.e. random variables that each can take only two values, 0 or 1; here, $Y_i = 1$ if the i-th individual tests positive for the disease and 0 otherwise. In this case, U_i cannot be included in the linear predictor, because *there is no such thing as extra-Bernoulli variation*. The explanation of this is the following. Firstly, suppose that Y_i given U_i takes the values 1 or 0 with probabilities $f(U_i)$ and $1 - f(U_i)$, respectively, for some function $f(\cdot)$. Then, unconditionally, $P(Y_i = 1) = E[f(U_i)] = p_i$, say. Now, because $U_1,\ldots,U_n$ are independent, and the Y_i are independent conditional on U_i, so are $Y_1,..,Y_n$, i.e. the Y_i are a set of independent Bernoulli trials, whatever the function $f(\cdot)$ and whatever the distribution of the U_i. See the text-box for a more formal explanation of this. It does not follow automatically that in the more general setting when explanatory variables $d(x_i)$ are introduced, the algebraic form of the relationship between $P(Y_i = 1)$ and $d(x)$ is preserved, but it is easy to show that this is the case for the logistic model (5.6).

In contrast, the inclusion of $S(x_i)$ in (5.6) when the response at each location is binary *can* be justified. In this case, although the Y_i are still marginally distributed as Bernoulli variables, the joint probability of a set of binary outcomes Y_i depends on the properties of the process S and, unlike those of U_i, these are statistically identifiable from binary data. We shall return to a discussion of binary outcomes in Section 5.6.

5.1.2 Poisson sampling

The Poisson distribution is widely used as a model for the sampling distribution of an outcome Y that is an open-ended count. A theoretical justification for this is that the Poisson distribution with mean $\mu = np$ approximates a binomial distribution in which the number of trials, n, is large and the probability of success, p, is small. In epidemiology this is an appropriate assumption for the number of cases of a rare disease in a large population, over as fixed spatial region and/or a fixed time-interval. In this context, the Poisson distribution can be derived directly by assuming that cases occur independently in a spatial or temporal continuum; see Section 5.6, where we discuss the link between geostatistical models and point processes.

Formally, the Poisson log-linear geostatistical model assumes that random variables Y_i associated with locations x_i are a set of independent Poisson random variables conditional on a zero-mean Gaussian process $S(x)$ and a set of independent zero-mean Normally distributed random variables U_i, with conditional expectations $\lambda(x_i)$ where

$$\log\{\lambda(x_i)\} = d(x_i)^\top \beta + S(x_i) + U_i. \tag{5.7}$$

The log-link in (5.7), as with the logit link for the binomial model, is widely used but not the only option.

A characteristic property of the Poisson distribution is that its variance and mean are equal. There is no guarantee that open-ended count data will satisfy this, and it is often the case in practice that, even under controlled conditions, the variance of an open-ended count is bigger than its mean. The non-spatial version of (5.7), i.e. without the term $S(x_i)$, is one way of accommodating this. Specifically, if Y conditional on U is Poisson-distributed with mean $\exp(\alpha + U) = \exp(\alpha) \times \exp(U)$, and U is Normally distributed with expectation $-\tau^2/2$ and variance τ^2 so that $E[\exp(U)] = 1$ whatever the value of τ^2, then unconditionally, Y has expectation $\mu = \exp(\alpha)$ and variance $\mu[1 + \mu\{\exp(\tau^2) - 1)\}] > \mu$ whenever $\tau^2 > 0$.

The inclusion of U_i in is also one of the possible approaches to account for an excess of zero counts, a problem also known as *zero-inflation*. This is illustrated in Figure 5.1. There, we consider the simple case of (5.7) in which $d(x)$ and $S(x)$ are both absent and we have re-scaled the U_i so that

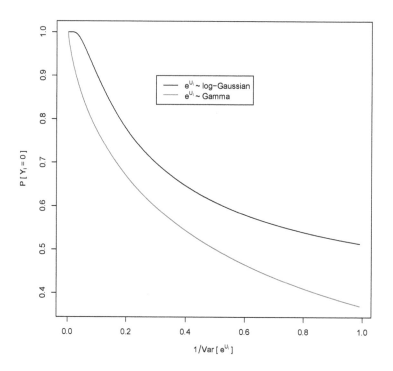

FIGURE 5.1
Plot of the marginal probability that $Y_i = 0$ from model (5.7) against $1/\operatorname{Var}[e_i^U]$, when e^{U_i} is log-Gaussian (black line) or Gamma (red line) distributed.

$E[Y_i] = E[\exp\{U_i\}] = 1$ whatever the variance of U_i. When $\operatorname{Var}[U_i] = 0$, Y_i is Poisson distributed with mean 1, and $P(Y_i = 0) = \exp(-1) \approx 0.37$. As $\operatorname{Var}[U_i]$ increases, $P[Y_i = 0]$ also increases towards a limiting value of 1. Formally, as $\operatorname{Var}[U_i] \to +\infty$, Y_i tends in distribution to a degenerate random variable with a probability mass of 1 at zero (although, paradoxically, $E[Y_i] = 1$ for any finite value of $\operatorname{Var}[U_i]$). In Chapter 8, we shall discuss the problem of zero-inflation in more detail and introduce more complex modelling approaches for cases where zero-inflation is a spatially structured phenomenon, and not simply a consequence of independent random extra-Poisson variation.

5.1.3 Negative binomial sampling?

Notice that the non-spatial version of (5.7) can equally be written as

$$\lambda(x_i) = W_i \exp\{d(x_i)^\top \beta\}, \tag{5.8}$$

where $W_i = \exp(U_i)$ follows a log-Gaussian distribution with expectation 1. If, instead, W_i is assigned a Gamma distribution with expectation 1 and shape parameter κ, the unconditional distribution of Y_i is negative binomial. In this case, if μ_i denotes the unconditional expectation and variance of Y_i, then its unconditional variance is $\mu_i(1 + \mu_i/\kappa)$.

Notice that, necessarily, whether W_i in (5.8) follows a log-Gaussian or a Gamma distribution does not alter the relationship between the unconditional expectation and variance of Y_i. This makes it difficult in practice to discriminate between these two distributional assumptions. (Firth, 1988).

Before the development of computational tools that made generalized linear mixed models routinely available to practising statisticians, the negative binomial distribution had the attraction of being an analytically tractable model for over-dispersed, open-ended count data and it is still widely used for this purpose. What is also sometime done is to use the spatial model (5.7) with both the random effect terms $S(x_i)$ and U_i and, at the same time, to assume that the conditional distribution of Y_i is negative binomial. In our opinion, this is not a good idea because it is very difficult to disentangle the effects of the negative binomial paremter κ and the variance of the U_i. For these reasons, we will assume throughout this chapter that the independent random effects U_i in (5.7) are Normally distributed.

5.2 Inference

5.2.1 Likelihood-based inference

The likelihood function for the vector of parameters θ in a GLGM is obtained by integrating out S from (4.4). However, in contrast to the linear model, this integral is no longer available in closed form. Here, we describe two strategies for evaluating the integral: the Laplace method, which uses a deterministic numerical approximation; and a Monte Carlo sampling method.

5.2.1.1 Laplace approximation

The Laplace method for approximating the likelihood function of a GLGM is based on a second-order Taylor expansion of the log-likelihood as follows. Write S as a shorthand for $[S, y; \theta]$, $\hat{S}$ for the maximised value of S at any particular value of θ,

$$G(S) = \frac{\partial \log[S, y; \theta]}{\partial S}$$

and

$$H(S) = -\frac{\partial^2 \log[S, y; \theta]}{\partial^2 S}.$$

Write the likelihood function as

$$L(\theta) = \int \exp\{\log[S, y; \theta]\} \, dS \tag{5.9}$$

The second-order Taylor expansion of $\log[S, y; \theta]$ about the point $\hat{S}$ is

$$
\begin{aligned}
\log[S, y; \theta] &\approx \log[\hat{S}, y; \theta] + (S - \hat{S})^\top G(\hat{S}) - \frac{1}{2}(S - \hat{S})^\top H(\hat{S})\,(S - \hat{S}) \\
&= \log(\hat{S}) - \frac{1}{2}(S - \hat{S})^\top H(\hat{S})\,(S - \hat{S}),
\end{aligned}
$$

where we have used the fact that $G(\hat{S}) = 0$. Substituting this last expression into (5.9) gives

$$
\begin{aligned}
L(\theta) &\approx \int \exp\{\log[\hat{S}, y; \theta] - 0.5(S - \hat{S})^\top H(\hat{S})\,(S - \hat{S})\} \, dS \\
&= \exp\{\log[\hat{S}, y; \theta]\} \int \exp\{-0.5(S - \hat{S})^\top H(\hat{S})\,(S - \hat{S})\} \, dS \\
&= \exp\{\log[\hat{S}, y; \theta])\} \times |H(\hat{S})|^{0.5} \times (2\pi)^{n/2} \times \\
&\quad \int (2\pi)^{-n/2} |H(\hat{S})|^{-0.5} \exp\{-0.5(S - \hat{S})^\top H(\hat{S})\,(S - \hat{S})\} \, dS
\end{aligned}
\tag{5.10}
$$

The integrand on the right-hand side of (5.10) is the probability density of a multivariate Normal distribution, hence the integral takes the value 1 and we obtain the Laplace approximation,

$$\log L_{LM}(\theta) = \log[\hat{S}, y; \theta] - \frac{1}{2}|H(\hat{S})|. \tag{5.11}$$

To estimate θ, we maximize (5.11) with respect to θ using numerical optimization. Confidence intervals for individual parameters can then be computed either by computing the inverse of the negative of the second derivative of $\log L_{LM}(\theta)$ evaluated at $\hat{\theta}$ and extracting the standard errors as the square roots of its diagonal elements, or through inspection of the profile likelihoods for each parameter.

The Laplace method is generally more accurate when $[y|S; \theta]$ can be well approximated by a Gaussian distribution. For a Binomial distribution, this holds when n_i is large and $p(x_i)$ is close neither to 0 nor to 1. For a Poisson distribution, the accuracy improves as $\lambda(x_i)$ increases. Joe (2008) is a study of the accuracy of the Laplace method for parameter estimation in generalized linear mixed models. The results show, as expected, that the method performs poorly when applied to binary data.

5.2.1.2   Monte Carlo maximum likelihood

An alternative approach to evaluating the integral (4.4) is to use a Monte Carlo method based on importance sampling, as follows. Let θ_0 denote any particular value of θ and write the likelihood function as

$$
\begin{aligned}
L(\theta) &= \int [S, y; \theta] \, dS \\
&= \int \frac{[S, y; \theta]}{[S, y; \theta_0]} [S, Y; \theta_0] \, dS \\
&\propto \int \frac{[S, y; \theta]}{[S, y; \theta_0]} [S|y; \theta_0] \, dS \\
&= E\left[\frac{[S, y; \theta]}{[S, y; \theta_0]}\right],
\end{aligned}
\tag{5.12}
$$

where the expectation is taken with respect to $[S|y; \theta_0]$. It follows that we can obtain an unbiased estimate of $L(\theta)$ for any value of θ by simulating B independent realisations of S conditional on y with θ held fixed at θ_0, say $s^{(j)} : j = 1, ..., B$, and taking their sample average,

$$
L_{MC}(\theta) = \frac{1}{B} \sum_{j=1}^{B} \frac{[s^{(j)}, y; \theta]}{[s^{(j)}, y; \theta_0]},
\tag{5.13}
$$

as an approximation to $L(\theta)$. Maximization of (5.13) gives the Monte Carlo maximum likelihood estimate $\hat{\theta}_{MC}$, which converges to the maximum likelihood estimate $\hat{\theta}$ as $B \to \infty$.

The above theory is quite general, but its successful implementation relies on several considerations. Most importantly, the quality of the Monte Carlo approximation improves the closer θ_0 is to the maximum likelihood estimate $\hat{\theta}$. In practice, this means choosing θ_0 to be our best guess at the value of $\hat{\theta}$; it is also both legitimate and advisable to obtain a provisional estimate of θ by maximising (5.13) and then repeat the maximisation with θ_0 set at this provisional estimate. Secondly, the computational efficiency of the optimization step can be improved by providing analytical expressions for the first and second derivatives of $L_{MC}(\theta)$. This is also helpful for exploring the surface of the likelihood function, especially for the covariance parameters, with respect to which the surface is often relatively flat. Finally, simulation of S conditional on y requires the use Markov Chain Monte Carlo algorithms, whose convergence must be trusted before using the simulated realisations to evaluate $L_{MC}(\theta)$. Section A.4 of the appendix gives more details.

5.2.2 Bayesian inference

In the previous section, we split the problem of estimating the parameters of a GLGM into two stages: calculation of the approximate likelihood at a fixed value of θ, using either a Laplace approximation or a Monte Carlo method; and optimisation of the likelihood. In Bayesian inference, the optimisation is replaced by an integration to convert the likelihood and prior to a posterior for θ. But because Bayeisan inference treats both the random effects, S and/or

U, and the parameters, θ, as unobserved random variables with specified distributions, it is natural to combine the two steps into a single algorithm.

In Bayesian statistics, a refinement of the Laplace approximation can be obtained through the Integrated Nested Laplace Approximation (INLA) methodology. This was introduced by Rue et al. (2009) and forms the basis of the widely used R-INLA software (http://www.r-inla.org/). INLA can produce very accurate approximations more quickly than Monte Carlo methods but, as noted in Section 5.2.1.1, is not a panacea. For example, Fong et al. (2010) report a poor performance of the INLA method in the case of binary data.

One of several possible Monte Carlo strategies for Bayesan inference in a GLGM is the following Markov Chain Monte Carlo (MCMC) algorithm, each iteration of which is broken down into three steps.

1) *Updating the covariance parameters.* This is often the most computationally intensive step in each iteration, as it requires the computation, inversion and determinant calculation of the covariance matrix for S. When vague priors are used, more iterations are likely to be required to obtain good mixing of the samples drawn from the Markov chain. To make the computations faster, Diggle & Ribeiro (2007) specified discrete priors for the covariance parameters, to allow pre-computation of each of the possible values of all the quantities needed for the implementation of the MCMC. If this requires more computer memory than is available, an alternative strategy is to use approximations of the spatial process S; see Section 3.6.

2) *Updating the regression coefficients.* The implementation of this step depends on the specific parametrization of the model. Let $\eta = (\eta_1, \ldots, \eta_n)$ where η_i is the linear predictor in (5.3). A convenient reparametrisation is obtained by the following factorisation of a GLGM

$$[\theta, \beta, \eta, Y] = [\theta][\beta|\theta][\eta|\beta, \theta][Y|\eta]. \qquad (5.14)$$

If $[\beta|\theta]$ is multivariate Gaussian, it can be shown that the full conditional distribution $[\beta|\theta, \eta, Y]$ is also Gaussian, in which case β can be updated using a Gibbs sampler. However, if multiple observations are available at a location x_i, e.g. multiple individuals within a household each with explanatory variable $d_j(x_i)$, the linear predictor is then expressed as

$$\eta_{ij} = d_j(x_i)^\top \beta + S(x_i) + U_i.$$

In such cases, the factorisation (5.14) is not convenient as some of its components will be perfectly correlated. An alternative option is to use the original parametrisation in (5.1) in conjunction with a random walk Metropolis-Hastings algorithm to update β. The PrevMap package uses an independence sampler in which, at each iteration, a new value for β is proposed from a Gaussian distribution with mean and covariance function given by the mode and inverse of the negative Hessian of the full conditional, respectively.

3) *Updating the random effects.* Several solutions have been proposed to implement this step; see, for example, Christensen et al. (2006). In `PrevMap`, the sum of the random effects is updated using a Hamiltonian Monte Carlo algorithm (Neal, 2011); for more details, we refer the reader to Section 2.2 of Giorgi & Diggle (2017).

5.3   Model validation

In Section 4.3, we set out a procedure for validating a fitted linear geostatistical model based on the empirical variogram of the residuals from an ordinary least squares fit to the fixed effects component of the model. We now suggest an analogous procedure for validation of a fitted GLGM.

The rationale for the procedure is that in a GLGM the variation in the outcome, over and above its intrinsic sampling variation, is partitioned into explained and unexplained variation *on the scale of the linear predictor*, and this is therefore the scale on which we want to assess whether, and if so how, the unexplained component is spatially correlated. Conventionally defined regression residuals, such as Pearson's residuals, do not achieve this. Instead, we proceed as follows.

We first fit to our data a non-spatial generalized linear mixed model (GLMM) with the same fixed effects as the model to be validated but independent random effects, i.e. a linear predictor of the form

$$\eta_i = d(x_i)^\top \beta + U_i,$$

where the U_i are assumed to be independent, Normally distributed with common mean zero and variance τ^2. We then calculate predicted values for the U_i. These could be any suitable summary of their corresponding predictive distributions, for example the mean or the mode. In the remainder of this chapter, we use the mode, corresponding to the value of U_i that maximizes the density function of its distribution conditional on the data, formally expressed as $[U_i; \tilde{\theta}][y_i|U_i; \tilde{\theta}]$, where $\tilde{\theta}$ is the estimated unknown vector of parameters under the GLMM. We then generalise the procedure described in Section 4.3 for the linear geostatistical model as follows. Firstly, calculate $\hat{V}_0(u)$, the variogram based on the $\hat{U}_i$. As in Section 4.3, fix the model parameters θ at their maximum likelihood estimate under the geostatistical model, or at some posterior point estimate if Bayesian methods are used, and proceed as follows.

1. Simulate the Gaussian process $S(x_i)$ and the Gaussian noise U_i at each of the observed locations $x_i : i = 1, \ldots, n$.

2. Simulate $Y_i : i = 1, ..., n$ as independent realisations of a generalized linear model with linear predictor $\eta_i = d(x_i)^\top \beta + S(x_i) + U_i$.

3. Compute the $\hat{U}_i := 1, ..., n$ from a spatially uncorrelated GLMM fitted to the simulated Y_i using explanatory variables $d(x_i)$.

4. Compute the variogram, $\hat{V}_1(u)$, based on $\hat{U}_i$.

5. Repeat steps 1 to 4 a large number of times, say B, to give variograms $\hat{V}_b(u) : b = 1, ..., B$.

Finally, use the resulting $\hat{V}_b(u)$ exactly as described in Section 4.3.

5.4   Spatial prediction

When carrying out spatial prediction, the first step is to define the predictive target, which we denote by T^*. Typically, this will be some property of the realisation of the spatial component of the linear predictor, i.e. the set of values of $d(x)^\top \beta + S(x)$ for all values of x in the region of interest A. However, according to the practical context, we will sometime want to focus on the unexplained component of the spatial variation, $S(x)$. For example, extreme values in a predictive map of $S(x)$ identify *anomalies* (sometime called "hot-spots" and "cold-spots") that might be of interest in their own right, or could provide clues to the nature of an omitted explanatory variable.

In the case of a geostatistical Binomial model (Section 5.1.1), a natural predictive target is the prevalence surface over the region of interest A, hence

$$T^* = \{p(x) = \exp\{T(x)\}/(1 + \exp\{T(x)\}) : x \in A\}, \qquad (5.15)$$

where $T(x) = d(x)^\top \beta + S(x)$. For a geostatistical log-linear Poisson model (Section 5.1.2) the analogous target would be the rate of occurrence,

$$\mathcal{P}(T^*) = \{\lambda(x) = \exp\{T(x)\} : x \in A\}. \qquad (5.16)$$

In both (5.15) and (5.16) a point prediction takes the form of a map. In any one application, the focus of interest, and hence the predictive target, may be not so much the map itself but rather a particular feature of the map; for example, its maximum, or an indicator of whether the average value over a sub-region exceeds a policy-relevant threshold. In such cases, it cannot be over-emphasised that simply calculating the property of interest from the predicted map is the wrong thing to do. As we now explain, the correct procedure is first to draw a number, B, of random samples from the predictive distribution of the complete spatial surface $\{S(x) : x \in A\}$, calculate the value of the specific target from each sample, say $T_1^*, ..., T_B^*$, and report suitable summaries of the resulting empirical distribution of the T_i^*.

Recall from Section 4.4 that, in order to make spatial prediction computationally feasible, we approximate A using a regular grid $\mathcal{X} = \{x_1^*, ..., x_q^*\}$ consisting of q prediction locations that cover A. To make inference on T^*, we need to obtain samples from its predictive distribution, $[T^*|y]$. Since any target T^* can be calculated directly from the fitted model parameters and the spatial surface $S(x)$, the problem reduces to sampling from the predictive distribution of $S^* = \{S(x) : x \in \mathcal{X}\}$.

Note that

$$[S^*|y] = \int [S^*, S|y] \, dS = \int [S|y][S^*|S] \, dS,$$

where we have used the fact that $[S^*|S, y] = [S^*|S]$. Hence, to sample from $[S^*|y]$, we need first to sample from $[S|y]$, and then from $[S^*|S]$ to obtain our sample s_h^*, for $h = 1, \ldots, B$. Also, note that $[S^*|S]$ a multivariate Gaussian distribution with mean vector and covariance matrix given by (4.14) and (4.15), respectively, but with y replaced by S. Hence, drawing samples from $[S^*|S]$ is straightforward.

Predictive samples s_h^* can then be transformed into corresponding samples t_h^* from the predictive distribution of T^* by direct calculation, and used to obtain any required summary of this predictive distribution, for example means, standard deviations or selected quantiles, at any or all of the q prediction locations x_j^*.

When the predictive target relates to the complete surface rather than a single summary, two ways to display uncertainty in the predictions are through quantile or exceedance probability surfaces. We define the α-*quantile surface* of any spatial process $W(x)$ as

$$\mathcal{Q}_\alpha(T^*) = \{q(x) : \mathrm{P}[W(x) < q(x)|y] = \alpha, x \in A\}. \tag{5.17}$$

Similarly, we define the c-*exceedance probability surface*,

$$\mathcal{R}_l(T^*) = \{r(x) : \mathrm{P}[W(x) > c|y] = r(x), x \in A\}. \tag{5.18}$$

Values of the point-wise exceedance probability $r(x_j^*)$ close to 1 identify locations for which is highly likely to exceed c, and vice-versa.

The choice of the process $W(x)$ is context-dependent. For example, when using the binomial geostatistical model to analyse prevalence survey data, the predictive target might be the spatial variation in the linear predictor, in which case $W(x) = T(x)$, or the prevalence itself, in which case $W(x) = g^{-1}\{T(x)\}$ where $g^{-1}\{\cdot\}$ is the inverse link function of the GLGM.

5.5 Applications

5.5.1 River-blindness in Liberia (continued)

In Section 2.2.1, we analysed the nodule prevalence data from Liberia using a non-spatial binomial regression model. Our exploratory analysis of the residuals from this non-spatial model suggested the presence of residual spatial correlation. To account for this, we now fit a spatial model in which, conditional on a Gaussian process $S(x_i)$, the numbers of people having nodules, Y_i, are mutually independent Binomial variables, with numbers of trials n_i, corresponding to the total number of tested people, and probabilities of nodule presence $p(x_i)$, such that

$$\log\left\{\frac{p(x_i)}{1 - p(x_i)}\right\} = \beta_0 + \beta_1 x_{i,1} + \beta_2 x_{i,2} + S(x_i), \qquad (5.19)$$

where $x_{i,1}$ and $x_{i,2}$ are, respectively, the east-west and north-south coordinates of the village location x_i. We specify $S(x)$ to be a stationary and isotropic zero-mean Gaussian process with variance σ^2 and exponential correlation function with scale parameter ϕ.

In Table 5.1, we compare the maximum likelihood estimates obtained from the Laplace and Monte Carlo maximum likelihood methods, with the latter using 5,000 simulations to calculate the Monte Carlo approximation of the likelihood function. The corresponding pairs of point estimates of all parameters differ only slightly, but the Laplace method delivers wider confidence intervals than the Monte Carlo approach. We view the Laplace approximation as a useful way of setting initial values for the Monte Carlo maximum likelihood algorithm in order to reach convergence more quickly.

TABLE 5.1

Maximum likelihood estimates and associates 95% confidence intervals (CI) for the model in (5.19) using the Laplace (Section 5.2.1.1) and the Monte Carlo likelihood (Section 5.2.1.2) methods.

Parameter	Laplace method Estimate	95% CI	Monte Carlo likelihood Estimate	95% CI
β_0	-6.284	(-9.461, -3.107)	-6.309	(-9.068, -3.550)
β_1	2.726	(-0.408, 5.860)	2.756	(0.264, 5.248)
β_2	4.751	(2.115, 7.386)	4.763	(2.207, 7.320)
$\log(\sigma^2)$	-1.946	(-3.025, -0.866)	-1.941	(-2.955, -0.926)
$\log(\phi)$	4.213	(2.966, 5.461)	4.221	(3.011, 5.431)

The graphical validation test for the assumed exponential correlation function of $S(x)$ is shown in Figure 5.2. The 95% tolerance envelope fully contains the empirical variogram based on the estimated random effects from a non-spatial binomial mixed model. Hence, we conclude that the data do not show evidence against the fitted correlation function. Note, however, that the simulation envelope is rather wide indicating, not untypically in our experience, that the data are limited in their capacity to discriminate amongst different families of spatial correlation function.

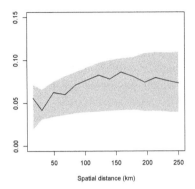

FIGURE 5.2
Diagnostic results for the analysis on river-blindness in Liberia. The solid line is the empirical variogram based on estimated random effects from a non-spatial binomial mixed model. The shaded area is a 95% tolerance bandwidth generated by the algorithm of Section 5.3.

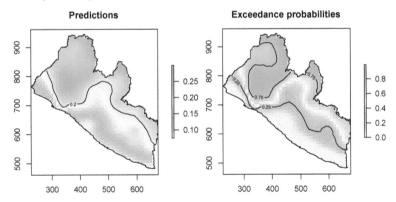

FIGURE 5.3
Maps of the predicted nodule prevalence (left panel) and exceedance probability of 20% prevalence (right panel) in Liberia. Contours of 20% prevalence and of 25% and 75% exceedance probabilities are shown in the left and right panels, respectively.

We now show the results for spatial prediction using a 10 by 10 km regular grid of prediction locations that cover the whole of Liberia. The left panel of Figure 5.3 shows that values of predicted nodule prevalence greater than 20% occur mainly in the north-western area of Liberia. The figure of 20% is relevant because this is the operational definition of a "high prevalence area," which affects the prioritisation of the area for a particular public health intervention. However, it would be statistically incorrect to consider these areas as truly exceeding 20% nodule prevalence because this would ignore the uncertainty in the spatial predictions of prevalence. In the right panel of Figure 5.3 we report the exceedance probability surface for a 20% prevalence threshold. The 75% contour on this map identifies a smaller proportion of north-west Liberia in which we are at least 75% confident that prevalence is greater than 20% The area between the 75% and 25% contours can be considered as a zone of uncertainty, in which additional data would need to be collected to decide whether or not these areas should be classified as high prevalence. More stringent exceedance probabilities, for example 95% and 5%, would necessarily extend the zone of uncertainty, illustrating the well-known statistical principle of "no free lunch."

5.5.2 Abundance of *Anopheles Gambiae* mosquitoes in Southern Cameroon (continued)

Our exploratory analysis on the abundance of *Anopheles gambiae* mosquitoes in Southern Cameroon (Section 2.2.2) hinted at the presence of residual small-scale spatial correlation, after accounting for the effect of elevation. To further investigate this aspect, we fit a geostatistical Poisson model in which, conditionally on a Gaussian process $S(x_i)$, the counts, Y_i, of mosquitoes at location x_i are mutually independent Poisson variables with means $\lambda(x_i)$ and linear predictors

$$\log\{\lambda(x_i)\} = \beta_0 + \beta_1 d(x_i) + S(x_i), \qquad (5.20)$$

where $d(x_i)$ is the elevation, in metres, at the location x_i. We again model the zero-mean Gaussian process $S(x)$ using an exponential covariance function.

TABLE 5.2
Monte Carlo maximum likelihood estimates and associated 95% confidence intervals for the model in (5.20).

Parameter	Estimate	95% CI
β_0	2.989	(1.698, 4.279)
$\beta_1 \times 10^3$	-2.254	(-4.234, -0.275)
$\log(\sigma^2)$	-0.314	(-0.669, 0.040)
$\log(\phi)$	0.443	(-0.748, 1.635)

Table 5.2 reports the Monte Carlo maximum likelihood estimates for the model (5.20). The point estimate of the scale parameter is $\hat{\phi} = 1.557$ km, indicating that the spatial correlation decays to a value of 0.05 at a distance of $\log 20 \times 1.558 \approx 4.7$ km, which is of the same order of magnitude as a mosquito's flight range, whereas the minimum observed distance between observations is 384 m. The model is therefore able to distinguish between spatial

and non-spatial components of extra-binomial variation, thereby strengthening the previously rather weak evidence for the existence of small-scale, but biologically plausible, residual spatial correlation. This also results in a larger confidence interval for β_1 than was reported in Table 2.3 from the non-spatial regression analysis.

5.6 A link between geostatistical models and point processes

In all of the geostatistical models that we have considered so far each response, Y, is associated with a point location, x. However, in many applications the recorded response is a property of a finite region for which x is simply a convenient reference location. For example, in Section 5.5.1, each Y refers to a village community that extends over a finite area, and x is a more-or-less arbitrary reference location somewhere within the village boundary. In classical geostatistics, the area from which Y is derived is called the *support* of Y. Informally, ignoring the support of each response is equivalent to assuming that for all quantities of interest, the variation within the extent of each single support is negligible compared with the variation between different supports. It follows that the larger the goegraphical extent of each support, the less plausible this assumption becomes.

In Section 5.5.1, each recorded response is a case-count and, in principle, each occurrence of the event of interest could have been allocated its own location, for example the place of residence of the individual concerned. This may not be feasible in practice, nor is it necessarily more appropriate epidemiologically than assigning a common location to all individuals within a single village. Nevertheless, it is of interest to consider how we would model data for which individual locations *are* available and relevant, not least because this will give some insight into the circumstances in which reducing a set of individual events to a count at a single nominal location could be misleading. Figure 5.4 gives a schematic representation of the problem. In the left-hand panel, individual events are shown as solid dots; in the right-hand panel, these are reduced to counts in each designated sampling area.

A statistical model for the locations of a set of events in a spatial region is a *spatial point process*. Models and statistical methods for analysing spatial point process data have been developed quite independently of geostatistical models and methods. Detailed accounts include Baddeley et al. (2016), Diggle (2013) or Ilian et al. (2008). However, there is a close connection between a particlar class of point process models and a cass of GLGMs.

The simplest spatial point process model is a *homogenous Poisson process*, in which the points are both independently and uniformly distributed at random over the area of interest. A more flexible class of models is an

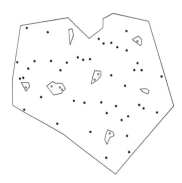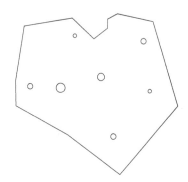

FIGURE 5.4
Schematic representation of individual events (solid dots) occurring in a geographical region. In the left-hand panel, events are coloured red or black
according to whether they do or do not fall within one of a set of delineated
sampling areas. In the right-hand panel, the number of events in each sampled area is shown as a circle at the centroid of each sampled area, with radius
size-coded to correspond to counts of 0, 1, 2, 3.

inhomogeneous Poisson process, in which the points are still distributed independently of each other, but with a probability density proportional to a
non-negative-valued function $\lambda(x)$, called the *intensity function* of the process;
in a homogeneous process, the intensity does not vary spatially, i.e. $\lambda(x) = \lambda$,
for all locations x. An important property of a Poisson process is that the
number of events in any spatial region A follows a Poisson distribution with
mean

$$\mu_A = \int_A \lambda(x)dx.$$

A second important property is that the numbers of events in non-overlapping
spatial regions are statistically independent.

Now consider how this point process model might be used in the scenario
shown in Figure 5.4. Suppose that the point process that generated the black
and red dots in the left-hand panel of Figure 5.4 is a Poisson process with
intensity $\lambda(x)$. The seven delineated small areas are non-overlapping, hence
the counts Y_i represented in the right-hand panel are independent and Poisson-
distributed with means

$$\lambda_i = \int_{A_i} \lambda(x)dx.$$

Now, suppose that the intensity function is a log-linear regression on a spatially varying covariate $d(x)$, so that

$$\log \lambda(x) = \alpha + \beta d(x). \tag{5.21}$$

Suppose also that the spatial variation in $d(x)$ is large-scale, in the sense that within any one of the seven delineated small areas $d(x)$ can be assigned a constant value, $d(x) = d(x_i)$ say, where x_i is any point in A_i, for example its centroid, without any serious loss of accuracy. It follows that

$$
\begin{aligned}
\lambda_i &= \int_{A_i} \exp\{\alpha + \beta d(x)\} dx \\
&\approx \int_{A_i} \exp\{\alpha + \beta d(x_i)\} dx \\
&= |A_i| \exp\{\alpha + \beta d(x_i)\} \\
&= \exp\{\alpha^* + \beta d(x_i)\}. \tag{5.22}
\end{aligned}
$$

The resulting model for the counts, Y_i, is simply a classical generalized linear model with log-link, Poisson error distribution and independent responses.

Returning to the point process setting, a Poisson process cannot capture dependence amongst the events of the process. One way to do so is to extend the model to allow the intensity function to be a real-valued stochastic process, rather than a deterministic function. A convenient way to do this is to add a stochastic term to the right-hand side of the log-linear regression (5.21), to give

$$\log \lambda(x) = \alpha + \beta d(x) + S(x), \tag{5.23}$$

where $S(x)$ is a Gaussian process. A point process model of this kind is called a *log-Gaussian Cox process* (Møller et al., 1998). The argument leading to (5.22) can now be repeated, leading to the conclusion that, conditional on the Gaussian process $S(x)$, the counts, Y_i, of the numbers of events in a set of non-overlapping small areas are independent and Poisson distributed with means

$$
\begin{aligned}
\lambda_i &= \int_{A_i} \exp\{\alpha + \beta d(x) + S(x)\} dx \tag{5.24} \\
&\approx \exp\{\alpha^* + \beta^* d(x_i) + S(x_i)\}, \tag{5.25}
\end{aligned}
$$

in which the closeness of the approximation depends on the extent to which neither the explanatory variable $d(x)$ nor the Gaussian process $S(x)$ vary substantially within any one of the small areas A_i.

It sometimes happens that case-locations are only available as counts, Y_i, of the numbers of cases in small areas A_i, but a relevant covariate is available as a raster image, i.e. in effect, a spatially continuous surface. It is them possible to fit the exact model (5.24) rather than the approximation (5.25). The details are beyond the scope of this book, but a description can be found in Diggle et al. (2013), and the methods are implemented in the R package `lgcp` (Taylor et al., 2013).

5.7 A link between geostatistical models and spatially discrete processes

In the previous section, we discussed a scenario in which a geostatistical data-set consists of counts of the numbers of events occurring in each of a set of small areas sampled from a much larger area. An alternative scenario, typical of disease registries in developed country settings, is when the data are again presened as counts in small areas, A_i say, but these form a partition of the whole of the study-area, A. The Spanish cancer atlas illustrated in Section 1.3 is a typical example.

In this situation, it is still the case that if the underlying point process of individual events is a log-Gaussian Cox process, then conditional on $S(x)$ the resulting counts are independent and Poisson-distributed with means given by (5.24). However, using the approximation (5.25), and analysing the data as if each small area were a single point is, at best, a questionable strategy. This is because the correlation between the spatial averages of a stationary process $S(s)$ over two contiguous subregions A_1 and A_2 depends materially on the sizes and shapes of A_1 and A_2 in addition to the distance between their centroids.

Most published methods for analysing of data of this kind side-step the question by taking the A_i as given, and modelling the responses $Y_i : i = 1, ..., n$ as a n n-dimensional random variable whose distribution is specific to this partition of A. The most widely used example, introduced by Besag et al. (1991), models the Y_i conditional on an unobserved multivariate Normal random variable $S = (S_1, ..., S_n)$ as independent Poisson random variables with means

$$\lambda_i = \exp\{\alpha + \beta d_i + S_i\} \tag{5.26}$$

where d_i is an explanatory variable associated with the small area A_i.

Equation (5.26) bears more than a passing resemblance to equation (5.25). The difference is that in (5.26) no attempt is made to give the model for S a spatially continuous interpretation. Instead, S is modelled a discrete *Markov random field* (Rue & Held, 2005), whose joint distribution is specified indirectly through its *full conditionals*, i.e. the n univariate conditional distributions of each S_i given all other S_j. If S is multivariate Normal, the full conditional distributions are univariate Normal, with means $\mu_i = a_i + \sum_{j \neq i} b_{ij} S_j$. In the current context, the usual way to give this model a spatial interpretation is to define the *neighbours* of A_i as all A_j that share a common boundary with A_i and set $b_{ij} = 0$ unless A_i and A_j are neighbours. Beyond this, a simple but remarkably effective special case has been used to construct many health atlases including the Spanish cancer atlas. This sets $a = 0$, $b_{ij} = 1/m_i$ where m_i is the number of neighbours of A_i so that the conditional mean μ_i is just the average of the values of neighbouring S_j, and specifies the conditional variances, σ_i^2 say, to be of the form $\sigma_i^2 = \phi/m_i$.

Diggle et al. (2013) show how to fit the exact spatially continuous log-Gaussian Cox process model (5.24) to data of this kind, with an illustrative application to the Spanish cancer data. The differences in the results obtained by the two approaches are small in that example, but become more pronounced when relevant explanatory variables are available at a finer spatial resolution than the small areas A_i and vary materially within the A_i. A separate consideration when formulating markov randon field models is whether the near-universal practice of defining small areas to be neighbours if they share a common boundary is reasonable. It is intuitively natural for spatially regular geographies, less so when small areas vary substantially in size and shape.

6

Geostatistical design

CONTENTS

6.1   Introduction

The planning, or *design*, of any scientific investigation involves many considerations, most obviously in the current context the number of measurement locations. However, one consideration that is specific to geospatial investigations, and is therefore the focus of this chapter, is *where* to take measurements, rather than simply how many. Another pragmatic reason for focusing on this aspect is that the number of measurement locations is usually cost-constrained, whereas it is often the case that different spatial arrangements of a fixed number of measurement locations will be cost-neutral.

Initially, we shall assume that the investigator is free to record the value of the outcome of interest at any location x within a geographical region A. Later, in describing search algorithms for optimal designs, we will approximate A by a finely spaced regular grid of potential sampling locations. In applications, the set of potential sampling locations may be further restricted by practical considerations to a particular finite set. For example, in designing a prevalence survey, measurement locations are necessarily restricted to inhabited locations within the region of interest.

We write $\mathcal{X} = \{x_1, ..., x_n\}$ for the set of n locations at which the outcome variable will be recorded and call $\mathcal{X}$ the *sampling design*. As we have seen in earlier chapters, geostatistical analysis mainly addresses either or both of two broad scientific objectives. The first is *estimation* of the parameters that define the stochastic model for the data $Y = \{y_i : i = 1, ..., n\}$ consisting of the recorded outcomes at each sampled location x_i. The second is *prediction* of the unobserved realisation of the unobserved process $S(x)$ throughout A, or particular aspects of this realisation that are of scientific interest. The fundamental geostatistical design problem is the specification of $\mathcal{X}$.

In the remainder of this chapter, for any explicit calculations that we report when comparing different designs, we will assume that the underlying model is a linear Gaussian model with no covariates, so that

$$Y_i = \mu + S(x_i) + Z_i : i = 1, ..., n, \tag{6.1}$$

where $S(x)$ is a stationary, Gaussian process with mean zero, variance σ^2 and correlation function $\rho(u)$, whilst the Z_i are mutually independent $N(0, \tau^2)$ random variables. Although this model is too restrictive for many applications, it is sufficient to illustrate the general principles of good geostatistical design.

In general, sampling designs that are efficient for parameter estimation may be inefficient for prediction, and vice versa (Zimmerman, 2006). In practice, most geostatistical problems focus on spatial prediction, but parameter estimation is an important means to this end. Hence, there is a need to compromise between designing for efficient parameter estimation and designing for efficient prediction given the values of relevant model parameters. We shall see that the spatial covariance structure of the process $S(x)$ plays a critical role in this compromise.

Ideally, a particular design $\mathcal{X}$ should be chosen to optimise a performance criterion that reflects the particular objective of the study; see, for example, Jardim & Ribeiro (2007) or Nowak (2010). For problems in which the primary goal is to predict the value of $S(x)$ throughout the study-region A, a conventional performance criterion is the spatially averaged mean squared prediction error,

$$MSPE = \int_A E[\{\hat{S}(x) - S(x)\}^2]dx, \tag{6.2}$$

where $\hat{S}(x) = E[S(x)|Y; \mathcal{X}]$ is the minimum mean square error predictor of $S(x)$ based on data Y and the expectation is with respect to the process $S(x)$. In any particular application, (6.2) may be of limited relevance because of the need to accommodate context-specific considerations. Foremost amongst these in low resource settings is cost, especially if travel to and between sample locations is time-consuming or expensive. There may also be purely scientific reasons why, in a particular application, precise prediction is more important in some parts of A than in others or, as was the case for the Liberian river blindness example discussed in Section 5.5.1, the important question is whether prevalence exceeds a policy-defined threshold for intervention, which in that example was set at 20%. A reasonable goal might then be to design the study so that optimal design is therefore one for which the predictive probability that prevalence exceeds 20% will be concentrated as closely as possible around either zero or one at all locations x. One of a number of possible design criteria that captures the spirit of this is

$$\int_A |\mathrm{Prob}\{P(x) > 0.2|Y\} - 0.5|dx.$$

6.2 Definitions

A sampling design $\mathcal{X}$ is *deterministic* if there is no element of randomness in its construction, otherwise it is *stochastic*. A stochastic design removes the potential for subjective bias in the choice of sampling locations x_i.

$\mathcal{X}$ is *uniform* if every location x in the region A is equally likely to be included in $\mathcal{X}$. In any particular context there may be good reason to over-sample some parts of A; for example, in pollution monitoring studies, monitors tend to be placed in areas where pollution is thought likely to be relatively high. Non-uniform designs do not necessarily lead to biased inferences, provided the design construction is documented and accounted for in the subsequent analysis of the data. But if the design is constructed haphazardly, without documentation, it can be difficult to correct for the resulting bias. We postpone further discussion of this until Chapter 7.

Our final distinction in this introductory section is between non-adaptive and adaptive designs. A design is *non-adaptive* if the complete design $\mathcal{X}$ is specified in advance of any data-collection. An *adaptive* design collects data in batches, and at each stage analysis of the available data can inform the design of the next batch.

6.3 Non-adaptive designs

The literature on designing for efficient parameter estimation includes work by Müller & Zimmerman (1999), Russo (1984) and Warrick & Myers (1987), who considered variogram-based estimation, and Lark (2002) and Pettitt & McBratney (1993) who considered likelihood-based methods. This body of work suggests that completely random designs are reasonably efficient for parameter estimation, and this has also been our experience. However, if the

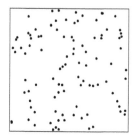 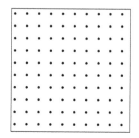

FIGURE 6.1

Completely random (left-hand panel) and square lattice (right-hand panel) designs, each consisting of $n = 100$ sampling locations on the unit square.

scientific focus is on prediction throughout a designated study-region $\mathcal{D}$ completely random designs seem less attractive as they can create relatively large unsampled patches within $\mathcal{D}$ (Müller, 2007).

A combination of theoretical and empirical evidence gathered over many years suggests that completely regular designs generally lead to efficient spatial prediction provided model parameters are known. A key reference is Matérn (1960, 1986). Others include McBratney et al. (1981), McBratney & Webster (1981), Yfantis et al. (1987), Ritter (1996) and Su & Cambanis (1993).

The assumption of a known covariance function is rarely realistic. In most applications we have to use the same data for estimation of covariance parameters and for spatial prediction. Articles that discuss designs for efficient spatial prediction in conjunction with parameter estimation include Diggle & Lophaven (2006), Zhu & Stein (2006), Banerjee et al. (2008), Bijleveld et al. (2012) and Chipeta et al. (2017).

6.3.1   Two extremes: completely random and completely regular designs

In a *completely random* design the locations x_i are an independent random sample from the uniform distribution on A. In a *completely regular* design, the x_i form a regular (typically square) lattice to cover A. Figure 6.1 shows an example of each. Both are of size $n = 100$. Both are stochastic (for the completely regular design the position of the bottom-left lattice point was chosen randomly), uniform and non-adaptive.

Classical statistical sampling theory emphasises the importance of random sampling as a protection against bias, but if the goal is to understand the spatial variation in some phenomenon $S(x)$ over the unit square, a completely random design is not the best choice. Given the opportunity to take measurements of $S(x)$ at a specified number of sampled locations, we suspect that most scientists would prefer to use a completely regular than a completely random design, and more often than not they would be right. An intuitive explanation for this is that most natural phenomena exhibit some degree of smoothness. Closely located pairs of locations are then likely to have similar values of x, and the second member of the pair therefore adds little information.

One theoretical limitation of a completely regular design is that the only element of randomness in the design is the choice of a sampling origin relative to which the lattice points are placed. Whilst this removes bias over repeated realisations of the design it is of scant comfort if, as is almost universally the case in practice, the design for a particular study consists of a single realisation.

If model parameters are unknown, another advantage of a completely random design over a completely regular design is that it will include a wider range of relatively small inter-point distances, which are particularly helpful for estimating the behaviour of the correlation function close to $u = 0$.

6.3.2 Inhibitory designs

An *inhibitory* design represents a compromise between a completely random and a completely regular design. Its construction uses the class of pairwise interaction point processes described in Ripley (1977). Here, we use the simplest form of pairwise interaction process in which there is a minimum permissible distance, d_0 say, between any two points but the process is otherwise completely random. The resulting designs exhibit a degree of spatial regularity without the deterministic geometry of a lattice design. For n points in a region A, the *packing density* of the design is $n\pi d_0^2/(4|A|)$, the fraction of the area of A occupied by n non-overlapping discs of diameter d_0. Figure 6.2 shows three examples with $n = 100$ and packing densities 0.1, 0.2 and 0.4. The progressive development of spatial regularity as the packing density increases is clear.

6.3.3 Inhibitory-plus-close-pairs designs

A critical consideration in geostatistical prediction is the behaviour of the variogram, $V(u)$, at small distances u. Recall from Chapter 3, equation (3.10) that the theoretical variogram of a stationary linear geostatistical model is

$$V(u) = \tau^2 + \sigma^2\{1 - \rho(u)\},$$

where $\rho(u)$ is the correlation function of the latent process $S(x)$, σ^2 is the variance of $S(x)$ and τ^2 is the conditional variance of a measurement, Y, at location x given the value of $S(x)$. Since $\rho(0) = 1$, it follows that $V(0) = \tau^2$. As discussed in Section 4.4, the predicted values of the surface $S(x)$ interpolate the data if $\tau^2 = 0$. Otherwise, they smooth out local fluctuations in the data

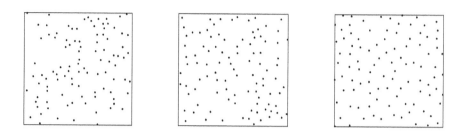

FIGURE 6.2
Three inhibitory designs with $n = 100$ points and packing densities 0.1, 0.2 and 0.4 (left, centre and right panels, respectively).

to an extent that depends on the ratio τ^2/σ^2. Consequently, if it is possible to obtain multiple, independent measurements of Y at the same location, this has the advantage that it allows direct estimation of τ^2. Specifically, if Y_1 and Y_2 are independent measurements at the same location, then $0.5(Y_1 - Y_2)^2$ has expectation τ^2 and the sample mean of all such quantities is an unbiased estimator for τ^2.

If the data do not contain multiple co-located measurements, estimation of τ^2 necessarily involves extrapolation, as illustrated in Figure 6.3. The diagram shows a hypothetical empirical variogram and two theoretical variograms that fit the empirical variogram equally well, but have substantially different values of τ^2 and would led to materially different predictions of $S(x)$, in two ways. Firstly, predictions obtained using the theoretical variogams indicated by the solid and dashed curves in Figure 6.3 would interpolate and smooth the data, respectively. Secondly, the predicted surface inherits the differentiability of the theoretical variogram at $u = 0$, hence in this hypothetical example the predicted surface corresponding to one of the two variograms (the one shown as a dashed curve in Figure 6.3) would be differentiable, the other not.

This leads us to consider a class of *inhibitory plus close pairs* designs, in which an inhibitory design is augmented by the inclusion of locations paired with, and in close proximity to, a sub-set of locations in the inhibitory design. A design in this class is defined by four quantities: n, the total number of points; k, the number of close pairs; d_0, the minimum distance between any two of the $n - k$ locations in the inhibitory component of the design; and r, the maximum distance between two paired locations. The exact placing of each paired location relative to its companion is drawn at random from a bivariate probability distribution; a suitable choice is the uniform distribution over the disc of radius r. Figure 6.4 shows two examples, each with $n = 100$

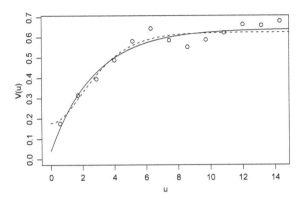

FIGURE 6.3
A hypothetical empirical variogam (open circles) and two theoretical variograms (solid and dashed lines) that give equally good fits.

points. The parameters of the two designs have been adjusted so that both have packing density 0.4 for their inhibitory component. The actual pairings are more obvious in the second example, where the two members of each pair are almost coincident.

Our experience has been that a relatively small number of close pairs, perhaps 10 or 20 in a data-set of size 100 or more, is sufficient. As a general rule of thumb, we also recommend that the packing density of the inhibitory points be in the range 0.4 to 0.5 when there is no prior restriction on the available sampling locations; a value as large as this may not be achievable otherwise.

The design construction can also be set up as a formal optimisation problem provided the user is willing to specify both a performance criterion and a hypothetical set of parameter values, say θ. Suppose, for example, that the number of measurement points, n, is fixed by constraints on time or cost and the goal is to minimise the spatially averaged mean squared error,

$$T = E\left[\int_{\mathcal{D}} \{\hat{S}(x) - S(x)\}^2 dx\right].$$

To approximate T, we first construct a set of prediction points, $\mathcal{X}_p = \{x_i^* : i = 1, ..., N\}$, as a finely spaced grid to cover $\mathcal{D}$, then calculate

$$\tilde{T} = N^{-1} \sum_{i=1}^{n} v(x_i^*), \qquad (6.3)$$

where $v(x)$ is the prediction variance, given by (4.18). Recall that $v(x)$ depends on the sampling design $\mathcal{X}$ and on the model parameters, θ, but not on the data, Y, whereas the function that we wish to optimise should depend only on the

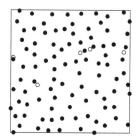 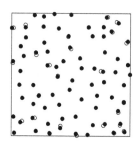

FIGURE 6.4
Two inhibitory plus close pairs designs with $n = 100$ points on the unit square. The design in the left-hand panel has 95 points in its inhibitory component and 5 paired points, each within a distance 0.05 of their companion. The design in the right-hand panel has 75 points in its inhibitory component and 25 paired points, each within a distance 0.02 of their companion. In both designs, the packing density of the inhibitory component is 0.4. Inhibitory and paired points are shown as closed and open circles, respectively.

design parameters, k, d_0 and r, i.e. our goal is to first to optimise a particular *class* of designs, then to generate a specific design within the optimum class. Hence, our design criterion, $\tilde{T}(k, d_0, r)$ say, should be the expectation of (6.3) over realisations of the design for specified values of k, d_0 and r. As the expectation is intractable it needs to be evaluated approximately as an average over simulated realisations of the design. Recall also that k is necessarily an integer. A more important limitation of optimised designs is that the necessary assumption of a known set of parameter values is seldom realistic.

6.3.3.1 Comparing designs: a simple example

Here, we illustrate the performance of three designs applied to realisations of two stationary Gaussian process with mean $\mu = 0$, variance, $\sigma^2 = 1$, nugget variance $\tau^2 = 0.2$ and exponential correlation function $\rho(u) = \exp(-u/\phi)$, with $\phi = 0.1$ (SGP1) and $\phi = 0.2$ (SGP2).

The three designs, each with $n = 100$ points on the unit square, are: D1 – a completely random design; D2 – a simple inhibitory design with packing density 0.5; D3 –an inhibitory-plus-close pairs design with 80 inhibitory points at packing density 0.5 and 20 close pairs, with distance 0.01 between the two members of a pair. See Figure 6.5.

 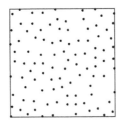 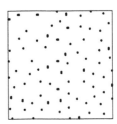

FIGURE 6.5
Three sampling designs with $n = 100$ points on the unit square: completely random (left-hand panel); inhibitory (centre panel); inhibitory plus close pairs (right-hand panel). See text for detailed specifications.

Figure 6.6 shows the realisations of the nugget-free spatial signal $S(x)$ that is our target for prediction, using a regular grid at a spacing of 0.01 to approximate the unit square. Note how the patches of positive and negative values of $S(x)$ tend to be larger in the realisation of SGP2 than in SGP1, as a consequence of its larger value of ϕ.

TABLE 6.1
Average squared prediction errors for three designs (D1, D2, D3) and realisations of two simulations models (SGP1, SGP2). See text for specification of designs and simulation models.

	D1	D2	D3
SGP1	0.606	0.507	0.539
SGP2	0.381	0.323	0.288

For each combination of design and model, Table 6.1 shows the sample mean of the squared prediction errors, $\{\hat{S}(x) - S(x)\}^2$, over the 10,000 points in a grid to cover the unit square at a spacing of 0.01. In each case, predictions were obtained by first sampling the value of Y_i at each of the 100 sample locations in D1, D2 or D3, then estimating all of the model parameters, μ, σ^2, ϕ and τ^2, by maximum likelihood, then calculating $\hat{S}(x_j)$ at each of the 10,000 grid-locations x_j as their conditional expectations given the data. Note that under either model, $Y_i = S(x_i) + Z_i$ where Z_i is distributed as $N(0, \tau^2)$ with $\tau^2 = 0.2$.

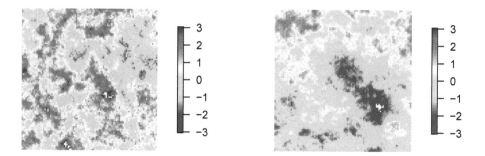

FIGURE 6.6

Realisations of two stationary process each with mean $\mu = 0$, variance $\sigma^2 = 1$ and exponential correlation function, $\rho(u) = \exp(-u/\phi)$. In the left-hand panel. $\phi = 0.1$. In the right-hand panel, $\phi = 0.2$.

In Table 6.1, the completely random design D1 gives the worst predictive performance amongst the three designs, both for SGP1 and SGP2. The lattice-plus-close pairs design gives the best performance for SGP2, but is beaten by the purely inhibitory design D2 for the nugget-free SGP1. Note also that prediction mean square errors are smaller for SGP2 than for SGP1, irrespective of the sampling design. This is a natural consequence of the fact that the correlation between the values of Y_i at any fixed data-location x_i and $S(x_j)$ at any fixed prediction location x_j increases as ϕ increases, all other things being equal. General conclusions should not be drawn from a single example, but the results are consistent with the view that the addition of close pairs to an inhibitory design is an effective, if potentially conservative, strategy when there is a need to estimate a nugget variance.

6.3.4 Modified regular lattice designs

Diggle & Lophaven (2006) propose and develop two different ways of augmenting a completely regular lattice design to guarantee the inclusion of closely spaced sample locations. The first of these adds close pairs in the same manner as for an inhibitory-plus close-pairs design. The second augments a primary lattice with more finely spaced in-fill lattices within each of a random sample of primary lattice cells. Figure 6.7 shows an example of each.

Diggle & Lophaven (2006) conclude that the close-pairs augmentation is generally preferable to the in-fill. We agree, and would also recommend the

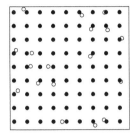 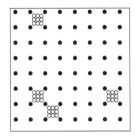

FIGURE 6.7
Two modified square lattice-based designs: lattice-plus-close-pairs (left-hand panel); lattice-plus-in-fill (right-hand panel). completely random (left-hand panel) and square lattice (right-hand panel) designs, each consists of $n = 100$ sampling locations on the unit square.

inhibitory-plus-close-pairs design over its lattice-based counterpart, primarily because of its greater flexibility.

6.3.5 Application: rolling malaria indicator survey sampling in the Majete perimeter, southern Malawi

In this section, we summarise an application from Chipeta et al. (2017) in which an inhibitory-plus-close-pairs design was used for a malaria prevalence survey in southern Malawi. The survey was an early component of an on-going (at the time of writing) five-year monitoring and evaluation study of malaria prevalence, with an embedded randomised trial of community-level interventions intended to reduce malaria transmission. The study area consists of three discrete administrative units, chosen from 19 such units that collectively surround the Majete Wildlife Reserve; see Figure 6.8, in which the three study-sub-areas are labelled A, B and C.

At an early stage of the investigation, the study team conducted a complete enumeration of households in each of the three sub-regions, including their geolocation using a GPS device with a notional accuracy of five metres. Using previously collected prevalence data from a sample of households in area B, Chipeta et al. (2017) fitted a Binomial logistic geostatistical model of the form (6.4), but with a simple intercept term for the fixed effects, hence

$$\log \left\{ \frac{p(x_i)}{1 - p(x_i)} \right\} = \alpha + S(x_i) + Z_i, \tag{6.4}$$

where the stationary process $S(x)$ has mean zero, variance σ^2 and correlation

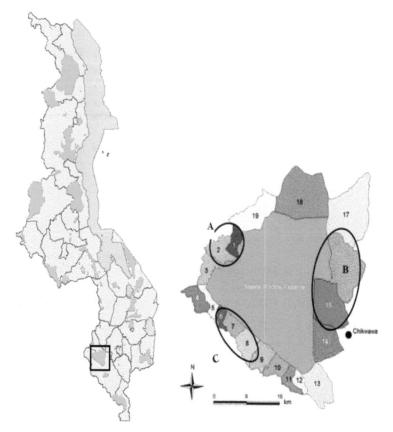

FIGURE 6.8
The map of Malawi, showing Majete Wildlife Reserve highlighted (left) and
its perimeter with focal areas A, B and C highlighted (right).

function $\rho(u) = \exp(-u/\phi)$, whilst the Z_i are mutually independent $N(0, \tau^2)$.
Table 6.2 summarises the results.

Chipeta et al. (2017) then used the parameter estimates from Table 6.2 to
optimize a sample of 200 households from the 857 candidates in sub-area A,
with optimisation criterion the sum of the mean squared prediction errors for
$S(x)$ at all 857 household locations,

$$SSPE = \sum_{i=1}^{857} E[\{\hat{S}(x_i) - S(x_i)\}^2].$$

Figure 6.9 shows the resulting design. As intended, the inhibitory component
of the design leads to good spatial coverage of the study sub-area within
the constraints imposed by the uneven spatial distribution of the households

Term	Estimate	95 % confidence interval
α	-1.910	(-2.190, -1.630)
σ^2	0.530	(0.318, 0.884)
τ^2	0.263	(0.074, 0.933)
ϕ	0.319	(0.133, 0.765)

TABLE 6.2
Monte Carlo maximum likelihood estimates and 95 % confidence intervals for the binomial logistic geostatistical model fitted to malaria prevalence data in Majete sub-area B.

over the sub-area, whilst the close pairs make for more precise parameter estimation.

6.4   Adaptive designs

This section draws heavily on Chipeta et al. (2016). Recall that an adaptive design collects data in batches and, at each stage, the analysis of the available data can inform the design of the next batch. One important characteristic of a design is its *batch size*. From a theoretical perspective the most efficient choice is *singleton adaptive sampling*, with batch size one. However, in most practical situations the batch size is constrained by operational constraints on the deployment of data-collectors including, for example, the costs associated with travelling between data-collection sites.

Our formal representation of an adaptive design is $\mathcal{X} = \{\mathcal{X}_0, \mathcal{X}_1, ..., \mathcal{X}_m\}$, where the suffix 0 denotes an *initial design* and remaining suffices denote batch numbers. We denote the corresponding outcome data by $Y = (Y_0, Y_1, ..., Y_m)$. Since later batches of outcome data Y_k are influenced by the locations in $\mathcal{X}_k$, which in turn are influenced by earlier batches of outcome data, there is a form of stochastic interplay between the $\mathcal{X}_k$ and the Y_k, and it is not immediately obvious how this might need to be taken into account in the subsequent analysis. In fact, the likelihood functions described in Chapters 4 and 5 remain valid under adaptive sampling. We postpone a proof of this until Chapter 7, where it arises naturally in a discussion of *preferential sampling* in geostatistics.

As with non-adaptive designs, the choice of performance criterion is of paramount importance. Suppose, temporarily, that singleton adaptive sampling is feasible. Given a criterion, $\mathcal{C}$ say, to be minimised, the optimum singleton addition to an existing but incomplete design is the currently unsampled location whose contribution to $\mathcal{C}$ is largest amongst the contributions of all available locations. For adaptive sampling with batch size $b > 1$, the naive

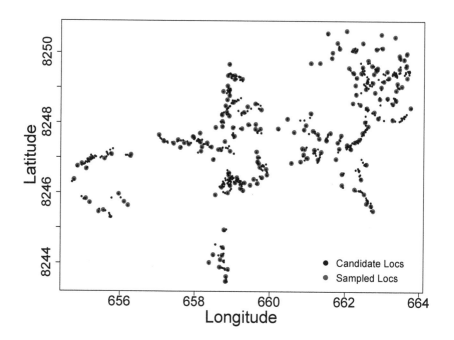

FIGURE 6.9
The optimised inhibitory plus close pairs design in sub-area A. Sampled and
unsampled households are indicated by blue and black dots, respectively

strategy of choosing the b available locations with the b largest current contributions to $\mathcal{C}$ is likely to fail. It typically selects a tight cluster of locations x
whose collective reduction in $\mathcal{C}$ may be little more than the reduction achieved
by any one of them, because the corresponding $S(x)$ are highly correlated,

6.4.1 An adaptive design algorithm

The first step in the design algorithm is to specify the complete set of N
potential sampling locations, $\mathcal{X}^* = \{x_1, ..., x_N\}$, in the study-region A. As
noted earlier, if every location in A could be sampled, in practice we specify
$\mathcal{X}^*$ to be a finely spaced grid of points to cover A.

The next step is to specify the size, n_0, of the initial design $\mathcal{X}_0$, for example
an inhibitory-plus-close-pairs design as recommended in Section 6.3, and the
number m and size b of subsequent batches $\mathcal{X}_k$ to constitute the final design
$\mathcal{X}$ of size $n = n_0 + mb$.

Thereafter, the algorithm proceeds as follows.

1. Obtain data Y_0 from the initial design $\mathcal{X}_0$ and fit a provisional model to these data.

2. Compute, or sample from, the predictive distribution of $S(x)$ at every unsampled location x in $\mathcal{X}^* = \{x_1, ..., x_N\}$, and hence the contribution of each unsampled location to the performance criterion T.

3. Select the unsampled location whose contribution to T is largest.

4. If $b > 1$:

    (a) select a second unsampled location whose contribution to T is largest subject to the constraint that it lies at least a prespecified distance d_0 from any location already included in the sample;

    (b) repeat until b locations have been identifed to form the set $\mathcal{X}_1$.

5. Obtain data Y_1 at the new location(s), return to 2 with the augmented design and data, and continue the cycle until n locations have been sampled.

An alternative stopping rule would be to specify an acceptable value of the T, and to continue sampling until a suitable summary of the predictive distribution of T, for example its prediction variance, reaches an acceptable level. The gains in efficiency that can be obtained by an adaptive sampling strategy will vary according to context. They can depend at least as much on practical as on theoretical considerations. The most important theoretical consideration in constructing an optimal non-adaptive design is whether the assumed model is at least approximately correct. A corresponding theoretical advantage of an adaptive strategy is that an initial design that, with hindsight, has been poorly chosen can be repaired progressively by the better-informed designs of later batches. Chipeta et al. (2016) gave a numerical illustration of the potential gains in a particular scenario. They assumed that the true surface was a realisation of a stationary Gaussian process on the unit square, with a Matérn correlation function and parameters $\phi = 0.05$, $\kappa = 1.5$ and $\tau^2 = 0$, i.e. no measurement error. Their performance criterion was the spatially averaged prediction variance, calculated numerically over a 64 by 64 grid; recall from results in Chapter 3 that this quantity depends only on the assumed model and the sample design, not on the actual values of $S(x)$, and that in the absence of measurement error, inhibitory designs are reasonably efficient. Their initial design was therefore a mildly inhibitory design, for which they chose a minimum distance $d_0 = 0.03$ between sampled locations, corresponding to a packing density of approximately 0.07. They then considered designs with total size $n = 100$, initial sizes $n_0 = 30, 40, ..., 100$ (the last of these being the comparator non-adaptive design) and batch sizes $b = 1, 5$ and 10. Figure 6.10 summarises the results. Two notable features are the very substantial gains

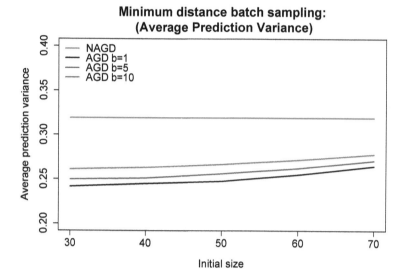

FIGURE 6.10
An efficiency comparison between non-adaptive (NAGD) and adaptive (AGD) designs with respect to spatially averaged prediction variance. All designs are inhibitory with minimum permissible distance between sampled locations $d_0 = 0.03$. For the adaptive designs, initial designs were of size $n_0 = 30, 40, ..., 90$, batch sizes were $b = 1, 5, 10$. Calculations assumed a Gaussian model with Matérn correlation and parameters $\phi = 0.05$, $\kappa = 1.5$. Adapted from Chipeta et al. (2016).

in efficiency for the smaller values of n_0 and the relatively small differences amongst the efficiencies obtained with the different batch sizes.

6.5   Application: sampling for malaria prevalence in the Majete perimeter (continued)

Recall from Section 6.3.5 that one aim of the Majete study is to monitor malaria prevalence in three sub-areas around the Majete wildlife reserve in southern Malawi. An initial malaria indicator survey conducted over the period April to June 2015 in sub-area B used an inhibitory sampling design with minimum distance constraint $d_0 = 0.03$ km and size $n = 100$ households, amongst which 72 households included at least one individual who met the inclusion criteria for the study.

Two covariates that were available on a raster scan covering the study-area, and which proved useful for predicting the spatial variation in malaria prevalance, were height above sea-level in metres (ELEV) the normalised digital vegetation index (NDVI). Both showed positive marginal associations with malaria, and a negative interaction. Chipeta et al. (2016) fitted a binomial logistic geostatistical model with exponential correlation function to the data from the three initial designs, obtaining the Monte Carlo maximum likelihood parameter estimates and 95% confidence intervals shown in Table 6.3.

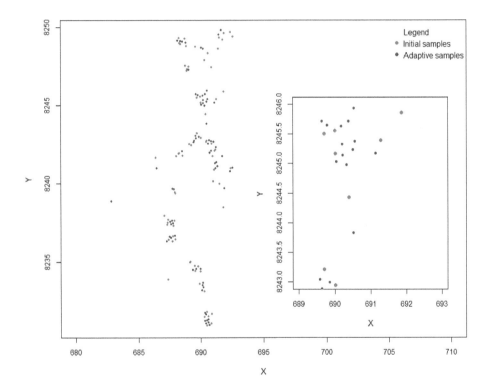

FIGURE 6.11
Initial inhibitory sampling design (red dots) and first wave of adaptive sampling locations (blue dots) in sub-area B. Inset shows an expanded view of a sub-set of sampled locations.

The overall study protocol specified that monthly follow-up prevalence surveys would be conducted, under the operational constraint that the field-team would be able to sample approximately 90 households per month. Hence the immediate design task was to tell the team which additional 90 households

they should visit. The result, using as performance criterion the spatial average
of the prediction variance of log-odds prevalence, is shown in Figure 6.11.

TABLE 6.3
Monte Carlo maximum likelihood estimates and 95 % confidence intervals for
the model fitted to the initial Majete malaria data.

Term	Estimate	95% Confidence Interval
Intercept	-5.483	(-7.6760, -3.2893)
ELEV	0.027	(0.0162, 0.0368)
NDVI	4.613	(0.1581, 9.0680)
Interaction	-0.040	(-0.0588, -0.0223)
σ^2	0.634	(0.4438, 0.9055)
ϕ	0.229	(0.1042, 0.5049)

6.6 Discussion

In any particular application, the objectives of the study can and should inform the design strategy. Our main aim in this chapter has been to describe a
framework for balancing the mildly competing considerations in designing for
efficient parameter estimation and for efficient prediction. In the latter case,
using spatially averaged mean square prediction error as a performance criterion provides a contextually neutral benchmark, but the underlying principles
hold for any context-specific performance criterion.

All of our recommended designs incorporate the widely accepted view
that spatial prediction is improved by using a more-regular-than-completely-
random configuration of sampling locations. As already noted, in our experience of evaluating inhibitory-plus-close-pairs design, a small proportion of
close pairs is sufficient. Moreover, substantial benefits from including close
pairs are only realised when the noise-to-signal ratio, τ^2/σ^2, of the best-fitting
model is relatively large.

7

Preferential sampling

CONTENTS

7.1 Definitions

The term *preferential sampling* was coined by Diggle et al. (2010) to mean that the process that generates the sampling locations x_i is stochastically dependent on the spatial process $S(x)$ that is of scientific interest. For reasons that will become clear, we shall call this *weakly preferential* sampling and add a stricter definition, namely *strongly preferential sampling*, to mean that the sampling locations and the latent process remain stochastically dependent conditionally on the measurements Y_i. It follows that, in the terminology of Chapter 6, a design can only be preferential if it is stochastic. Also, both non-adaptive and adaptive desgins may or may not be preferential.

Strongly preferential sampling as here defined is the geostatistical counterpart of *informative follow-up* in longitudinal studies whereby, for example in a medical setting, a patient presents for measurement when they are experiencing a particular combination of symptoms; see, for example, Lin et al. (2004).

A concept that is closely related to strongly preferential sampling is *informative missingness*, meaning that the *absence* of a measurement at a particular location x conveys information about the value of $S(x)$. The analogous problem in the context of longitudinal data has been studied extensively; see, for example, Diggle & Kenward (1994), Little (1995), Hogan & Laird (1997), Daniels & Hogan (2008). Missingness can only be identified when the set of intended sampling locations is specified beforehand. This is a common scenario in the longitudinal setting, where a study-protocol will typically specify a schedule of follow-up times at each of which it is intended to record the outcome of interest for every study-participant. In the authors' experience it is less common in the geospatial setting, which may simply be because its occurrence is not documented. For example, if a longitudinal study protocol specifies that the outcome of interest is measured at times $t = 0, 1, 2, 3$ and an individual study-participant provide data only at $t = 0, 1, 3$, the absence of a datum at $t = 2$ will be obvious in any presentation of the data, but if a geostatistical study protocol specifies that a single mesasurement should be made at each of n locations and in the event data are only gathered at $n - 1$ of these, the data may be presented as if there was never any intention to use the missing location, thereby assuming implicitly that missingness is *non-informative*. This seems to be what has happened in the case of a study of lead pollution in Galicia, north-west Spain, that we will analyse in Section 7.3. In the right-hand panel of Figure 7.1, it appears that the intention was to use a completely regular lattice design but, for unknown reasons, one or two lattice points are missing and several other sampling locations are perturbed from their corresponding lattice points.

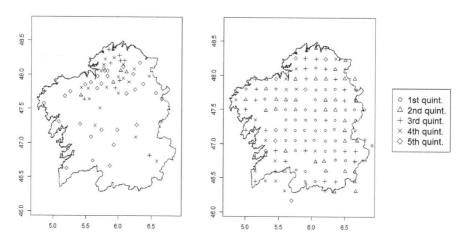

FIGURE 7.1
Sampling locations for the two surveys of lead concentrations in moss samples. The two maps correspond to surveys conducted in 1997 (left panel) and 2000 (right panel). Unit of distance is 100 km. Each point is represented by a symbol corresponding to a quintile class of the observed lead-concentration values as indicated by the legend.

In the left-hand panel of Figure 7.1 the sampling desgin is clearly non-uniform with a higher intensity of sampling in the northern part of Galicia and may or may not be preferential.

In environmental monitoring applications, a common reason for a measurement to be recorded as missing is that the value of whatever is being measured falls below a detection limit. In this case, provided that the detection limit is a known constant, the data are not strictly missing, but rather are *right-censored* and should be analysed accordingly, although this may present technical difficulties.

7.2 Preferential sampling methodology

Recall our general notation $\mathcal{S}$, $\mathcal{X}$ and $\mathcal{Y}$ to denote the signal process, the design and the measurement data, respectively. As always, we use $[\cdot]$ to mean "the distribution of $\cdot$", and a vertical bar to denote stochastic conditioning.

A general factorisation of the joint distribution of $\mathcal{S}$, $\mathcal{X}$ and $\mathcal{Y}$ is

$$[\mathcal{S}, \mathcal{X}, \mathcal{Y}] = [\mathcal{S}][\mathcal{X}|\mathcal{S}][\mathcal{Y}|\mathcal{X}, \mathcal{S}], \tag{7.1}$$

and integration with respect to $\mathcal{S}$ gives the likelihood function based on the data $\mathcal{X}$ and $\mathcal{Y}$. Other factorisations are, of course, available but (7.1) is the most natural from a modelling perspective because the signal process precedes the design and, in the non-adaptive case, the design precedes the measurement data. Integration of (7.1) with respect to $\mathcal{S}$ gives the log-likelihood function for the data, $\mathcal{X}$ and $\mathcal{Y}$, as

$$L(\mathcal{X}, \mathcal{Y}) = \log \int [\mathcal{X}|\mathcal{S}][\mathcal{Y}|\mathcal{X}, \mathcal{S}][\mathcal{S}]d\mathcal{S}. \tag{7.2}$$

Under non-preferential sampling, $[\mathcal{X}|\mathcal{S}] = [\mathcal{X}]$, hence (7.2) becomes

$$L(\mathcal{X}, \mathcal{Y}) = \log \int [\mathcal{Y}|\mathcal{X}, \mathcal{S}][\mathcal{S}]d\mathcal{S} + \log[\mathcal{X}]. \tag{7.3}$$

It follows from (7.3) that the stochastic variation in $\mathcal{X}$ can be ignored for inference about $\mathcal{S}$ or $\mathcal{Y}$. Conventional geostatistical methods do precisely this, by treating the design as a fixed set of locations $x_i : i = 1, ..., n$ and, typically, assuming that the measurements Y_i are conditionally independent given the corresponding $S(x_i)$, hence $[\mathcal{Y}|\mathcal{X}, \mathcal{S}] = \prod_{i=1}^{n}[Y_i|S(x_i)]$.

To explain why the distinction between weakly and strongly preferential sampling matters, we use an alternative factorisation to (7.1), namely

$$[\mathcal{S}, \mathcal{X}, \mathcal{Y}] = [\mathcal{S}][\mathcal{Y}|\mathcal{S}][\mathcal{X}|\mathcal{Y}, \mathcal{S}], \tag{7.4}$$

Under weakly preferential sampling, $[\mathcal{X}|\mathcal{Y}, \mathcal{S}] = [\mathcal{X}|\mathcal{Y}]$ and the log-likelihood becomes

$$L(\mathcal{X}, \mathcal{Y}) = \log \int [\mathcal{Y}|\mathcal{S}][\mathcal{S}]d\mathcal{S} + \log[\mathcal{X}|\mathcal{Y}]. \qquad (7.5)$$

Ignoring the term $\log[\mathcal{X}|\mathcal{Y}]$ in (7.5) therefore leads to inferences about $\mathcal{S}$ or $\mathcal{Y}$ that are valid, but may or may not be inefficient depending on the exact specification of $[\mathcal{X}|\mathcal{Y}]$. Under strongly preferential sampling, no factorisation of $[\mathcal{S}, \mathcal{X}, \mathcal{Y}]$ separates $\mathcal{X}$ and $\mathcal{S}$, hence valid inference requires the stochastic nature of $\mathcal{X}$ to be taken into account, i.e. $\mathcal{X}$ is not ignorable.

In the remainder of this chapter, we use the unqualified term *preferential* as a shorthand for *strongly preferential*

7.2.1 Non-uniform designs need not be preferential

It is easy to imagine circumstances in which a non-uniform spatial distribution of locations has practical advantages. These include: stratification of the study-region, with relatively high sampling intensity in sub-regions of particular concern (for example, monitoring air pollution more intensively in areas of high population density); maximising overall efficiency when some sub-regions are relatively inaccessible and therefore cost more to sample; designing to span the range of a potentially important covariate.

All of these circumstances can be captured by adding to our notation $\mathcal{D}$, to denote the set of all design factors. Then, a design is non-preferential if

$$[\mathcal{X}|\mathcal{S}; \mathcal{D}] = [\mathcal{X}; \mathcal{D}]. \qquad (7.6)$$

In (7.6), we make a notational distinction between conditional dependence on a stochastic quantity, denoted by the vertical bar, and functional dependence on a non-stochastic quantity, denoted by the semi-colon.

7.2.2 Adaptive designs need not be strongly preferential

The essence of adaptive design is to exploit the information about $\mathcal{S}$ gleaned from an initial set of sampling locations in order to maximise the additional information that can be gained from later sampling locations. It is intuitively clear that this must induce some form of dependence between $\mathcal{X}$ and $\mathcal{S}$ in the complete process. To explain why adaptive designs need not be strongly preferential, let $\mathcal{X}_0$ denote an initial sampling design chosen independently of $\mathcal{S}$, and $\mathcal{Y}_0$ the resulting measurement data. Analysis of the resulting data informs the choice of a further batch of sampling locations $\mathcal{X}_1$, which generate additional data $\mathcal{Y}_1$, and so on. The complete data-set consists of $\mathcal{X} = \{\mathcal{X}_0, \mathcal{X}_1, ...\mathcal{X}_k\}$ and $\mathcal{Y} = \{\mathcal{Y}_0, \mathcal{Y}_1, ..., \mathcal{Y}_k\}$. The associated likelihood for the complete data-set is

$$[\mathcal{X}; \mathcal{Y}] = \int [\mathcal{X}; \mathcal{Y}; \mathcal{S}]d\mathcal{S} \qquad (7.7)$$

We consider first the case $k = 1$. The standard factorisation of any multivariate distribution gives

$$[\mathcal{X}, \mathcal{Y}, \mathcal{S}] = [\mathcal{S}, X_0, Y_0, X_1, Y_1] \quad = \quad [\mathcal{S}][\mathcal{X}_0|\mathcal{S}][\mathcal{Y}_0|\mathcal{X}_0, \mathcal{S}][\mathcal{X}_1|\mathcal{Y}_0, \mathcal{X}_0, \mathcal{S}] \times$$
$$[\mathcal{Y}_1|\mathcal{X}_1, \mathcal{Y}_0, \mathcal{X}_0, \mathcal{S}]. \tag{7.8}$$

On the right-hand side of (7.8), note that by construction, $[\mathcal{X}_0|\mathcal{S}] = [\mathcal{X}_0]$ and $[\mathcal{X}_1|\mathcal{Y}_0, \mathcal{X}_0, \mathcal{S}] = [\mathcal{X}_1|\mathcal{Y}_0, \mathcal{X}_0]$. It then follows from (7.7) and (7.8) that

$$[\mathcal{X}, \mathcal{Y}] = [\mathcal{X}_0][\mathcal{X}_1|\mathcal{Y}_0, \mathcal{X}_0] \int [\mathcal{Y}_0|\mathcal{X}_0, \mathcal{S}][\mathcal{Y}_1|\mathcal{X}_1, \mathcal{Y}_0, \mathcal{X}_0; \mathcal{S}][\mathcal{S}]d\mathcal{S}. \tag{7.9}$$

The term outside the integral in (7.9) is the conditional distribution of X given Y_0. Using the fact, again by construction, that $[\mathcal{Y}_0|\mathcal{X}_1, \mathcal{X}_0; \mathcal{S}] = [\mathcal{Y}_0|\mathcal{X}_0; \mathcal{S}]$ the integral simplifies to

$$\int [\mathcal{Y}|\mathcal{X}, \mathcal{S}][\mathcal{S}]d\mathcal{S} = [\mathcal{Y}|\mathcal{X}].$$

It follows that

$$[\mathcal{X}, \mathcal{Y}] = [\mathcal{X}|\mathcal{Y}_0][\mathcal{Y}|\mathcal{X}]. \tag{7.10}$$

Equation (7.10) shows that the conditional likelihood, $[\mathcal{Y}|\mathcal{X}]$, can legitimately be used for inference although, depending on how $[\mathcal{X}|\mathcal{Y}_0]$ is specified, it may be inefficient. The argument leading to (7.10) extends to $k > 1$ with essentially only notational changes.

7.2.3   The Diggle, Menezes and Su model

Diggle, Menezes and Su (2010, henceforth DMS) used a parametric model to demonstrate some of the practical implications of strongly preferential sampling. Specifically, they assumed that $\mathcal{S}$ is a stationary Gaussian process and that, conditional on $\mathcal{S}$, $\mathcal{X}$ is an inhomogeneous Poisson process with intensity $\Lambda(x) = \exp\{\alpha + \beta S(x)\}$. Unconditionally, $\mathcal{X}$ is a log-Gaussian Cox process (Møller et al., 1998). This corresponds to non-preferential sampling when $\beta = 0$, and to strongly preferential sampling otherwise. Fitting the Diggle et al. (2010) model to the lead pollution data illustrated in Figure 7.1 gave clear evidence of strongly preferential sampling. Also, allowance for this produced predictions of $\mathcal{S}$ that were materially different from predictions based on the standard linear Gaussian model assuming non-preferential sampling. However, the use of a single parameter to capture both the strength of the non-preferentiality and the amount of non-uniformity in $\mathcal{X}$ is somewhat inflexible. In the next section, we therefore describe a more flexible class of models.

7.2.4   The Pati, Reich and Dunson model

Pati, Reich and Dunson (2011, henceforth PRD) proposed an extension of the DMS model in which they added a second Gaussian process, as follows.

Firstly, we again assume that $\mathcal{X}$ is a log-Gaussian process with intensity

$$\Lambda(x) = \exp\{\alpha + S_1(x)\}, \tag{7.11}$$

where S_1 is a Gaussian process with mean zero, variance σ_1^2 and correlation function $\rho(u/\phi_1)$.

Secondly, we assume that measurements $Y_i : i = 1, ..., n$ at locations x_i follow the model

$$Y_i = \mu + \beta S_1(x_i) + S_2(x_i) + Z_i, \tag{7.12}$$

where S_2 is an isotropic Gaussian process, independent of S_1, with mean zero, variance σ_2^2 and correlation function $\rho(u/\phi_2)$, where u is the Euclidean distance between any two locations. Also, the Z_i are mutually independent $N(0, \tau^2)$ variates. The model reduces to a re-parameterised version of the DMS model when the process S_2 is absent, i.e. $\sigma_2^2 = 0$. Otherwise, for convenience we replace τ^2 by $\nu^2\sigma_2^2$, so that the parameter ν^2 represents a noise-to-signal ratio.

In the model defined by (7.11) and (7.12), the parameter β controls the degree of preferentiality in the sampling of the Y_i, whilst the process S_2 represents a component of the spatial variation in the measurement process that is not linked to the sampling process. This allows us to parameterise the model so that the preferentiality parameter β does not feature in the sub-model for the sampling intensity $\Lambda(x)$. As shown in the next section, this in turn allows us to develop an inferential algorithm that uses an approximation to the field S_1 to circumvent the otherwise intractable distribution of $\Lambda(x)$

7.2.4.1   Monte Carlo maximum likelihood using stochastic partial differential equations

We now outline a Monte Carlo maximum likelihood procedure for parameter estimation under the PRD model. Setting $\mathcal{S} = \{S_1, S_2\}$, we rewrite (7.2) as

$$
\begin{aligned}
[\mathcal{X}; \mathcal{Y}] &= \int\int [\mathcal{X}; \mathcal{Y}; S_1; S_2]\, dS_1 dS_2 \\
&= \int\int [S_1][S_2][\mathcal{X}|S_1][\mathcal{Y}|\mathcal{X}; \mathcal{S}]\, dS_1 dS_2. \tag{7.13}
\end{aligned}
$$

In the integrand on the right-hand-side of (7.13), the elements of $\mathcal{X}$ given S_1 are independently and identically distributed with probability density

$$\frac{\Lambda(x)}{\int \Lambda(u)\, du}, x \in A. \tag{7.14}$$

The integral that forms the denominator of (7.14) is intractable. For this reason, we use an approximation of S_1 .

PRD approximate S_1 using a low-rank approximation based on a kernel convolution representation (Higdon, 1998) (see also Section 3.6.1). However, this approach has substantial computational advantages only for relative large values of the scale of the spatial correlation parameter ϕ_1. Diggle et al. (2013) use an extended regular grid on which they define $\mathcal{S}_1$. By wrapping this grid onto a torus, they then carry out inversion of the spatial covariance matrix using Fourier methods.

Here, we approximate $\mathcal{S}_1$ using a technique based on stochastic partial differential equations, which is described in Section 3.6.2. The high computational efficiency and ease of implementation of this approach make it an attractive alternative to either low-rank approximations or Fourier methods.

Integration of (7.13) with respect to $\mathcal{S}_2$ is analytically tractable, and leads to the approximation

$$[\mathcal{X}; \mathcal{Y}] = \int [\mathcal{S}_1][\mathcal{X}|\mathcal{S}_1][\mathcal{Y}|\mathcal{X}; \mathcal{S}_1] \, d\mathcal{S}_1 \approx \int [\tilde{\mathcal{S}}_1][\mathcal{X}|\tilde{\mathcal{S}}_1][\mathcal{Y}|\mathcal{X}; \tilde{\mathcal{S}}_1] \, d\tilde{\mathcal{S}}_1. \quad (7.15)$$

Now, let θ denote the vector of parameters of $\tilde{\mathcal{S}}_1$, which we denote by writing $\tilde{\mathcal{S}}_1(\theta)$, and write (7.15) as

$$
\begin{aligned}
\int [\tilde{\mathcal{S}}_1(\theta); \mathcal{X}; \mathcal{Y}] \, d\tilde{\mathcal{S}}_1 &= \int \frac{[\tilde{\mathcal{S}}_1(\theta); \mathcal{X}; \mathcal{Y}]}{[\tilde{\mathcal{S}}_1(\theta_0); \mathcal{X}; \mathcal{Y}]} [\tilde{\mathcal{S}}_1(\theta_0); \mathcal{X}; \mathcal{Y}] \, d\tilde{\mathcal{S}}_1 \\
&\propto \int \frac{[\tilde{\mathcal{S}}_1(\theta); \mathcal{X}; \mathcal{Y}]}{[\tilde{\mathcal{S}}_1(\theta_0); \mathcal{X}; \mathcal{Y}]} [\tilde{\mathcal{S}}_1(\theta_0)|\mathcal{X}; \mathcal{Y}] \, d\tilde{\mathcal{S}}_1, \quad (7.16)
\end{aligned}
$$

where θ_0 represents a "best guess" for $\hat{\theta}$.

To compute the integral in (7.16), we then use an independence sampler (IS) to simulate N realisations from the conditional distribution $[\tilde{\mathcal{S}}_1(\theta_0)|\mathcal{X}; \mathcal{Y}]$. At each iteration of the IS algorithm, we propose a new value from a multivariate Student's t distribution with 10 degrees of freedom, and location parameter and dispersion matrix given by the mode of $[\tilde{\mathcal{S}}_1(\theta_0); \mathcal{X}; \mathcal{Y}]$ and the inverse of the negative Hessian at the mode, respectively. Let $\tilde{\mathcal{S}}_{1(j)}$ denote the j-th out of the N simulated samples. The final approximation of the likelihood function is then given by

$$\frac{1}{N} \sum_{j=1}^{N} \frac{[\tilde{\mathcal{S}}_{1(j)}(\theta); \mathcal{X}; \mathcal{Y}]}{[\tilde{\mathcal{S}}_{1(j)}(\theta_0); \mathcal{X}; \mathcal{Y}]}. \quad (7.17)$$

To maximize (7.17) with respect to θ, we use a Newton-Raphson algorithm based on analytical expressions of the gradient function and Hessian matrix. After obtaining the Monte Carlo maximum likelihood estimate, say $\hat{\theta}_N$, we reiterate the algorithm setting $\theta_0 = \hat{\theta}_N$ and repeat this procedure until convergence.

7.3   Lead pollution in Galicia

We now apply the PRD model of Section 7.2.4 to the data on lead concentrations in moss samples shown in Figure 7.1. The data derive from two surveys
of the same area (Galicia, north-west Spain), conducted in 1997 and 2000
(Figure 7.1). The 1997 survey used a non-uniform, and therefore potentially
preferential, sampling design with more intensive sampling in the northern
part of Galicia. The 2000 survey used an approximate lattice design that,
notwithstanding the minor reservations noted earlier, we assume to be non-
preferential, i.e. for the 2000 survey data, $\beta = 0$.

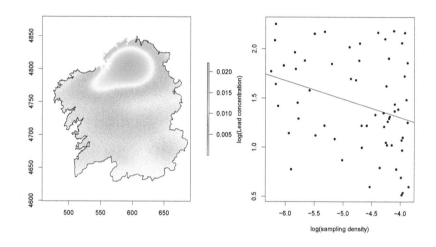

FIGURE 7.2
Kernel density estimate, $\hat{f}(x)$, of the sampling density for the 1997 lead concentration data (left-hand panel), scatter plot of log-transformed lead concentrations, Y_i, against $\log \hat{f}(x_i)$ (right-hand panel, dashed line is the least
squares fit.

The two panels of Figure 7.1 show the sampling locations of the 1997 and
2000 surveys. Each point in the map is represented by a symbol corresponding
to a quintile class of the observed lead-concentration values. We see that the
level of lead concentration was higher in 1997, with a North-South trend that
is not obviously present in the 2000 data.

A useful exploratory device to look for evidence of preferential sampling
is a scatterplot of the response variable against a non-parametric estimate of
the sampling density at each measurement location. The right-hand panel of
Figure 7.2 is an example of this plot for the 1997 log-transformed lead pollution data, with sampling density estimated using a simple kernel smoother,

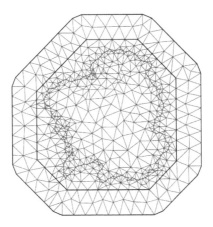

FIGURE 7.3
Triangulated mesh used for the SPDE representation of the field $\mathcal{S}_1$.

show in the left-hand panel. The plot shows a weak negative association, indicating a higher concentration of monitoring stations in areas with lower lead concentration. This in turn suggests preferential over-sampling of locations with lower-than-average pollution levels. Note, in this context, that Schlather et al. (2004) proposed a formal test that we could also have used here as an exploratory device.

Following DMS, we specify an exponential correlation function for $\mathcal{S}_2$. We use a Matérn covariance function with $\kappa = 1$ for $\mathcal{S}_1$ in order to have an explicit expression for the SPDE approximation. Figure 7.3 shows the triangulated mesh on which the SPDE representation for $\mathcal{S}_1$ is built. The triangles outside the borders of Galicia are used in order to avoid edge effects. We allow the mean parameter of the measurement process to differ between the two surveys, but assume that other model parameters are common to both; DMS provided some evidence to support this by fitting independent Gaussian processes to the 1997 and 2000 data, incorporating a nugget variance and an exponential correlation function, hence variogram

$$V(u) = \tau^2 + \sigma^2\{1 - \exp(-u/\phi)\}.$$

They fitted the model under four scenarios: non-preferential or preferential sampling in 1997 combined with separate parameters in 1997 and 2000 or common values of τ^2, σ^2 and ϕ in the two years. Table 7.1 shows the four maximised log-likelihoods. These provide strong evidence that the 1997 sampling is preferential, but weak evidence that for requiring separate values of the parameters τ^2, σ^2 and ϕ.

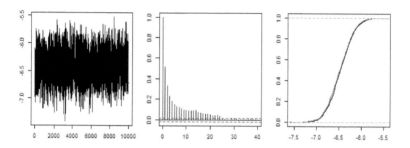

FIGURE 7.4
Trace plot (central panel), correlogram (central panel) and empirical cumulative density function (right panel) of the first and second 5000 samples of the region-wide average of $\tilde{S}_1$, obtained from the independence sampler algorithm.

TABLE 7.1
Maximised log-likelihoods for four models fitted by Diggle et al. (2010) to the Galicia lead pollution data.

1997 preferential?	common, τ^2, σ^2 and ϕ^2?	maximised log-likelihood
no	no	-96.79
	yes	100.62
yes	no	-82.95
	yes	-86.04

We now estimate the parameters of the PDR model using Monte Carlo maximum likelihood and combining the data from the 1997 and 2000 surveys. We simulate 10,000 samples from $[\mathcal{S}_1|\mathcal{X};\mathcal{Y}]$ by iterating the IS algorithm 12,000 times, with a burn-in of 2,000. Figure 7.4 shows some diagnostic plots on the convergence of the IS algorithm based on the region-wide average of $\tilde{S}_1$, given by

$$\frac{1}{|A|} \int_A \tilde{S}_1(x)\, dx$$

where A corresponds to the whole of Galicia and $|A|$ is its area. These indicate a satisfactory mixing of the Markov chain.

Table 7.2 shows Monte Carlo maximum likelihood estimates and associated 95% confidence intervals for the fitted model.

To the extent that they are directly comparable, the point estimates in Table 7.2 are qualitatively similar to those reported in DMS. Specifically, mean levels are lower in 2000 than in 1997 ($\mu_{00} < \mu_{97}$), and the estimate of the preferentiality parameter β is negative. However, the confidence interval

TABLE 7.2
Monte Carlo maximum likelihood parameter estimates and 95% confidence interval for the model fitted to the 1997 and 2000 surveys of lead concentrations in moss samples. The unit of measure for ϕ_1 and ϕ_2 is 100km.

Parameter	Estimate	95% confidence interval
α	-0.356	(-1.006, 0.293)
μ_{97}	0.854	(-0.565, 2.273)
μ_{00}	0.725	(0.527, 0.923)
β	-0.114	(-0.345, 0.117)
σ_1^2	6.610	(1.832, 23.822)
σ_2^2	0.148	(0.093, 0.237)
ϕ_1	1.855	(0.976, 3.525)
ϕ_2	0.235	(0.121, 0.459)
ν^2	0.349	(0.112, 1.089)

for β comfortably includes zero, whereas DMS found significant evidence of preferential sampling. An explanation for this apparent incompatibility can be seen in the values of the estimated spatial variance components, σ_1^2 and σ_2^2, which together with the estimate of β imply that most of the spatial variation in lead concentrations is unrelated to the spatial heterogeneity in the distribution of the 1997 sampling locations.

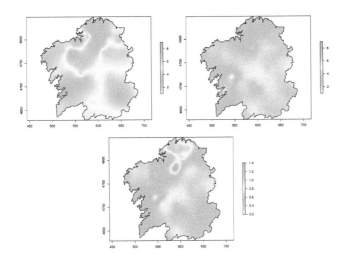

FIGURE 7.5
Plug-in predictions for the spatial variation in lead concentrations in 1997 (upper left panel) and 2000 (upper panel), and the ratio of the latter over the former (lower panel)

Our re-analysis using the PDR model also finds that the range of the spatial variation is much larger for $\mathcal{S}_1$ than for $\mathcal{S}_2$. The point estimate of ϕ_2 is comparable to that obtained by DMS. However, the additional flexibility of the PDR model has allowed us to capture the different spatial scales of variation in the relative sampling intensity and the lead concentration surface.

Figure 7.5 shows plug-in predictions for the lead concentration surfaces in 1997 and in 2000, and for the ratio of the two. As expected, the 1997 surface shows considerable spatial variation unrelated to the variation in sampling intensity shown in the left-hand panel of Figure 7.2, whilst the ratio of the 1997 and 2000 surfaces shows the spatial variation relative to the overall reduction in lead concentrations between the two dates with, however, also some increase in the northern area of Galicia.

Ozone pollution

- **What is ozone?** Ozone is gas that naturally occurs both in the atmosphere and at ground a level. A molecule of ozone consists of three atoms of oxygen (O_3).

- **Good ozone.** Ozone that naturally occurs in the upper layer of the atmosphere known as stratosphere forms a shield that protect us from the harm of ultraviolet rays. The so called *hole in the ozone* is the reduction of stratospheric ozone due to man-made chemicals.

- **Bad ozone.** Ozone present in the troposphere (the lowest layer of the earth's atmosphere) is the product of chemical reactions between pollutants emitted by cars, power plants, industrial boilers and other sources, with sunlight. Tropospheric ozone is also one of the main components of *smog*, i.e. smoke and fog created by air pollutants.

- **Source.** `www.epa.gov/ozone-pollution/basic-information-about-ozone`

7.4 Mapping ozone concentration in Eastern United States

We analyse data on ozone concentration, measured in part per billions (ppb), from Eastern United States. These data were originally described and analysed by Pati et al. (2011).

Figure 7.6 shows the spatial locations of the air pollution monitoring stations. We can see that these are not evenly distributed within the study area but form small clusters along the coast in some inland areas.

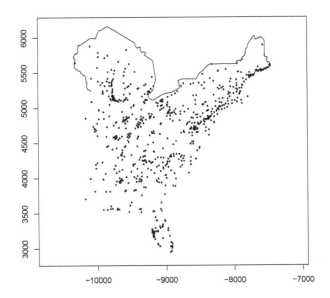

FIGURE 7.6
Locations of the air pollution monitoring stations across the Eastern United States.

The upper panel of Figure 7.7 shows the association between population density and the sampling intensity estimated using a smoothing Kernel. The *broken stick*, estimated using ordinary least squares with a knot at 5 for the logged population density, shows that as population increases the sampling intensity also increases and then levels off. This suggests that the monitoring stations have been mostly placed in populated areas which are expected to have higher levels of pollution. The middle and lower panels of Figure 7.7 indeed support this hypothesis.

Following from our exploratory analysis, we model the underlying sampling process as a log-Gaussian Cox process with log-intensity given by

$$\log\{\Lambda(x)\} = \gamma_0 + \gamma_1 d(x) + \gamma_2 \max(0, d(x) - 5) + S_1(x)$$

where $d(x)$ is the log-transformed population density after adding 1. The model for ozone concentration is

$$Y_i = \xi_0 + \xi_1 d(x_i) + \xi_2 \max(0, d(x_i) - 5) + \beta S_1(x_i) + S_2(x_i) + Z_i.$$

The estimates and 95% confidence intervals for the regression coefficients

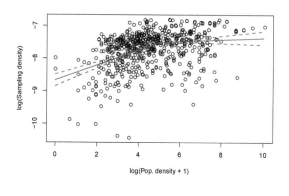

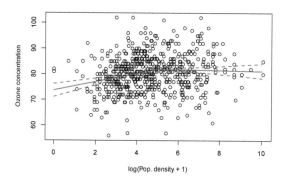

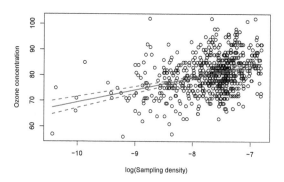

FIGURE 7.7

Scatter plots of the sampling intensity against the log-transformed population density (upper panel), the ozone concentration against the log-transformed population density (middle panel) and the ozone concentration against the sampling intensity. The sampling intensity is estimated using a Gaussian kernel density estimate. The red solid lines are least squares fit with corresponding 95% confidence intervals as dashed lines. The *broken sticks* in the upper and lower panels are constructed using a knot at 5 for the log-transformed population density.

(see Table 7.3) associated with populations density indicate that this has a significant effect both on the sampling intensity and the ozone concentration. However, the large positive estimate for the preferentiality parameter β suggests the presence of factors in addition to population density as contributing to the stochastic dependence between the sampling process and the spatial process for the ozone concentration.

TABLE 7.3
Monte Carlo maximum likelihood estimates and associated 95% confidence intervals for the model on ozone concentration.

Parameter	Estimate	95% CI
γ_0	-10.990	(-11.353, -10.627)
γ_1	0.769	(0.676, 0.862)
γ_2	-0.362	(-0.545, -0.180)
σ_1^2	0.708	(0.608, 0.825)
ϕ_1	175.295	(114.751, 267.783)
ξ_0	74.921	(70.844, 78.999)
ξ_1	0.797	(0.200, 1.394)
ξ_2	-0.941	(-1.932, 0.050)
σ_2^2	43.825	(16.821, 114.177)
ϕ_2	258.715	(136.088, 491.842)
τ^2	16.424	(9.825, 27.457)
β	2.586	(0.985, 4.188)

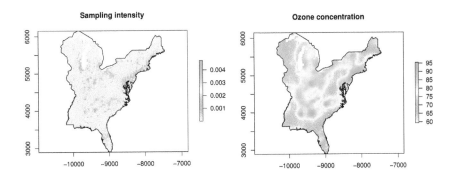

FIGURE 7.8
Predicted surfaces for the sampling intensity (left panel) and ozone concentration in ppb (right panel).

The left and right panels of Figure 7.8 show the maps of the predicted surfaces for the sampling intensity $\Lambda(x)$ and the ozone concentration across the Eastern United States. Higher levels of ozone are found along the coast and inland in places where the sampling intensity is also higher. However, the latter shows a patchier surface than that estimated surface for the ozone concentration. This is partly explained by the estimate of ϕ_2 which is about 47% larger than that for ϕ_1 (see Table 7.8).

7.5 Discussion

In any model for a preferentially sampled geostatistical data-set, at least one of the parameters is likely to be poorly identified. DMS already experienced this in their analysis of the Galicia lead pollution data, and the extra flexibility of the PRD model can only exacerbate the problem. Nevertheless, we would recommend using the PRD model as a vehicle for investigating preferential sampling effects, rather than the over-rigid DMS model.

In the analysis of the Galicia lead pollution data, our strategy for dealing with poor identifiability has followed that of DMS in arguing that the lattice-like sampling design of the 2000 survey can safely be assumed to be non-preferential, the models for the 1997 and 2000 pollution surfaces can also be assumed to have some parameters in common, and parameters can therefore be estimated by pooling the two contributions to the overall likelihood.

In general, several different strategies for investigating potentially preferential sampling can be considered. One would be to identify covariates that can partially explain the spatial variation in both the sampling intensity and the variable of interest. As in other areas of statistical modelling with random effects, we can think of the unobserved surface $\mathcal{S}_1$ that features in both (7.11) and (7.12) as a proxy for the combined effects of unmeasured covariates. As expressed formally by (7.6), addition of covariates to a model can render what would otherwise be preferential sampling non-preferential. The same applies *a fortiori* to spatially stratified sampling, provided that the stratification forms part of the original study-design.

A third strategy would be to treat the preferential sampling parameter, i.e. β in the PRD model, as a pre-specified constant and conduct a sensitivity analysis. We would recommend this when only a single data-set is available, with no obvious candidate covariates and no knowledge of how the sample locations were chosen.

A fourth would be to use Bayesian inference with an informative prior for the preferential sampling parameter. It is unclear to us how one might elicit this, and an uninformative prior would hide rather than solve the problem. Of course, this is not a reason for avoiding Bayesian methods if that is the analyst's preferred inferential paradigm.

Strongly preferential sampling invalidates conventional geostatistical inferences. Whenever possible it should be avoided by using explicit probability-based sampling designs. However, this is a counsel of perfection. When analysing observational geostatistical data, investigating whether preferential sampling effects are present is a better strategy than simply ignoring them.

Using a probability-based design should not equate to sampling completely at random with the study-region. As discussed in Chapter 6, in the absence of covariate information sampling designs that are spatially more regular than random generally lead to more precise predictions. When covariate information is available, spanning the range of important covariates will pay dividends in the same way that it does in simple regression modelling. Stratification of the study-region is also a useful tool, especially when precise prediction is more important in some sub-regions, for example in areas of high population density. Also as discussed in Chapter 6 adaptive designs, which are weakly preferential, can also bring gains in efficiency of prediction by comparison with analogous non-adaptive designs.

8

Zero-inflation

CONTENTS

The natural boundaries of a disease which encompass areas with suitable conditions for its transmission are often unknown. This can lead to over-sampling of communities that reside in areas where endemic transmission is either absent or at such low levels that the disease does not represent a public health threat. As a result of this, the data can contain an excess of zero reported cases, a phenomenon known as *zero-inflation* which has been studied for some tropical diseases; see, for example, Oluwole et al. (2015) for a case-study on soil-transmitted helminths and Amek et al. (2011) and Giardina et al. (2012) on malaria.

In this chapter, we first introduce a general modelling framework that deals with zero-inflation and discuss three main forms of spatially structured zero-inflation for disease counts data. We then described methods of inference and spatial prediction and, finally, illustrate two applications in river-blindness and Loa loa mapping.

8.1   Models with zero-inflation

For the sake of simplicity and without loss of generality, we describe different forms of zero-inflation by restricting our attention to disease prevalence data. More specifically, we consider three scenarios, as shown in Figure 8.1, based on different assumptions about the prevalence pattern as we approach the boundary between endemic and non-endemic areas: prevalence smoothly approaches but never reaches zero; prevalence smoothly approaches and exactly

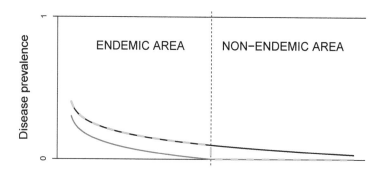

FIGURE 8.1
Plot of the three different disease prevalence patterns at the boundary of endemic areas: continuous and decreasing trajectory that never reaches exactly zero (black line); continuous and decreasing trajectory that exactly reaches zero (red line); and discontinuous trajectory that exactly reaches zero (green line).

reaches zero; and prevalence reaches exactly zero following a discontinuous trajectory.

Let Y_i be the random variable associated with the number of positively tested individuals out of n_i at location x_i, for $i = 1, \ldots, n$. The first form of zero-inflation, corresponding to the black line of Figure 8.1, can be obtained by modelling Y_i as a mixture of two variables, namely $Y_{1,i}$, which corresponds to the zero-inflation process, and $Y_{2,i}$, which models the counting process. More specifically, our model for the data can be expressed as

$$Y_i = Y_{1,i} Y_{2,i} \tag{8.1}$$

where $Y_{1,i}$ is a Bernoulli variable such that $P(Y_1 = 1) = \pi(x_i)$ and $Y_{2,i}$ follows a Binomial distribution with probability of success $p(x_i)$ and number of trials n_i. If we assume that $Y_{1,i}$ and $Y_{2,i}$ are independent of each other, it follows that

$$P(Y_i = 0) \quad = \quad 1 - \pi(x_i) + \pi(x_i)P(Y_{2,i} = 0).$$

It can be easily shown that the above quantity is always greater than or equal to $P(Y_{2,i} = 0)$, as long as

$$P(Y_{2,i} = 0) \leq 1$$

which is true by definition. This shows that the model for Y_i yields a greater probability that a 0 outcome will occur than a Binomial model.

Let $S(x)$, $T(x)$ and $U(x)$ denote three independent spatial Gaussian process. Diggle & Giorgi (2016) suggest to model $\pi(x)$ and $p(x)$ using a logit-linear regression, i.e.

$$\log\left\{\frac{p(x_i)}{1 - p(x_i)}\right\} = d(x_i)^\top\beta + S(x_i) + U(x_i)$$

and

$$\log\left\{\frac{\pi(x_i)}{1 - pi(x_i)}\right\} = d(x_i)^\top\gamma + T(x_i) + U(x_i)$$

where β and γ are vectors of regression coefficients. The residual process $U(x)$ is interpreted as shared spatial variation between the zero-inflation and counting processes. Diggle & Giorgi (2016) also note that a large amount of data would be required in order to accurately estimate all three Gaussian processes. They then suggest as a pragmatic to set $U(x) = 0$ for all x.

In model (8.1), disease prevalence is now defined by the product $\pi(x_i)p(x_i)$ which is always positive, hence it does not exclude the occurrence of cases in non-endemic areas. Whether this assumption is reasonable is context-specific. For example, we may not exclude the possibility of importation of cases from endemic areas to disease-free areas. If, instead, reintroduction of the disease in a non-endemic area is extremely unlikely, we may want to impose prevalence to exactly reach zero.

A second form of zero inflation, represented by the red line in Figure 8.1, is obtained using a thresholding approach. Conditionally on a Gaussian process $S(x)$, we now define the distribution of Y_i as

$$P(Y_i = y_i | S(x_i)) = \begin{cases} 0 & \text{if } d(x_i)^\top\beta + S(x_i) < 0 \\ \binom{n_i}{y_i}p(x_i)^{n_i}(1 - p(x_i))^{n_i - y_i} & \text{if } d(x_i)^\top\beta + S(x_i) > 0 \end{cases}$$

$$(8.2)$$

where

$$p(x_i) = 2\frac{\exp\{d(x_i)^\top\beta + S(x_i)\}}{1 + \exp\{d(x_i)^\top\beta + S(x_i)\}} - 1.$$

The above model defines non-endemic areas as the set of locations x such that $d(x_i)^\top\beta + S(x) < 0$. If x_0 is a point that lies on the boundary of such region, we also note that $p(x)$ would tend to zero as $\|x - x_0\| \to 0$, implying the continuity of disease prevalence at the boundary between endemic and non-endemic areas.

However, discontinuity in disease prevalence might arise from targeted control interventions and/or as a result of the urban-rural divide. This scenario is represented by the dashed, green line of Figure 8. Under this scenario, we could then model $Y_{1,i}$ as a binary spatial process and $Y_{2,i}$ as Binomial. The resulting model for Y_i is given by

$$P(Y_i = y_i | S(x_i), Y_{1,i}) = \begin{cases} 0 & \text{if } Y_{1,i} = 0 \\ \binom{n_i}{y_i}p(x_i)^{n_i}(1 - p(x_i))^{n_i - y_i} & \text{if } Y_{1,i} = 1 \end{cases} \quad (8.3)$$

where $\log\{p(x_i)/(1-p(x_i))\} = d(x_i)^\top \beta + S(x_i)$.

An example of a spatial binary process is the Ising model, named after the physicist Ernst Ising who originally proposed this a as a probabilistic model for ferromagnetism but it nowadays finds application in different fields of science, including ecology, agriculture and image analysis. Let $\tilde{x}_j$ denote the j-th location of a regular grid covering the region of interest. In its simplest form, the Ising model for $Y_{1,i} = Y_1(\tilde{x}_j)$ is defined as

$$P\left(Y_1(\tilde{x}_i) = 1 \middle| \sum_{j \sim i} I(Y_1(\tilde{x}_j) = 1) = k_i\right) \propto \exp\{\alpha k_i\},$$

where k_i is the number of neighbouring cells $\tilde{x}_j$ to $\tilde{x}_i$ such that $Y(\tilde{x}_j) = 1$ and α is a parameter that regulates the strength of the spatial interaction. For more details and extensions of the Ising model, we refer to Hurn et al. (2003).

In practice it is empirically difficult to reliably distinguish between the three outlined forms of zero-inflation, unless subject matter knowledge can help us to favour one over the others. In the remainder of this chapter, we shall focus our attention on the most widely used form of zero-inflation, which corresponds to (8.1).

8.2 Inference

We now provide more details on how to carry out likelihood-based inference for the zero-inflated model in (8.1). More generally, we let Y_2 follow any probability distribution, whether discrete or continuous.

Let $[\cdot]$ be a shorthand notation for "the distribution of $\cdot$". Let $y = (y_1, \ldots, y_n)$ be the observed realizations of $Y = (Y_1, \ldots, Y_n)$ at locations $\{x_1, \ldots, x_n\}$. Finally write $W = (S(x_1) + U(x_1), \ldots, S(x_n) + U(x_n), T(x_1) + U(x_1), \ldots, T(x_n) + U(x_n))$. The likelihood function for unknown vector of model parameters θ is then given by

$$L(\theta) = [y; \theta] = \int [W; \theta][y|W; \theta] \ dW \tag{8.4}$$

where $[W; \theta]$ is a multivariate Gaussian distribution with zero mean and co-variance function given by

$$\Sigma_W = \begin{bmatrix} \Sigma_S + \Sigma_U & \Sigma_U \\ \Sigma_U & \Sigma_S + \Sigma_U \end{bmatrix}$$

with Σ_S, Σ_T and Σ_U denoting the spatial covariance matrices of the Gaussian processes $S(x)$, $T(x)$ and $U(x)$, respectively. Finally, following from the assumption of conditional independence, the second factor of the integrand in

(8.4) has expression

$$[y|W] = \prod_{i=1}[y_i|W_i, W_{n+i}; \theta].$$

If we want to further expand the above equation and write it in terms of $Y_{1,i}$ and $Y_{2,i}$, which we leave as an exercise to the reader, we first need to specify whether $Y_{2,i}$ is continuous or discrete. Hence, we note that, if $Y_{2,i}$ is continuous, $P(Y_{2,i} = 0) = 0$ and, therefore, whenever $Y_{1,i} = 1$, this would imply that $y_i > 0$, whilst $y_i = 0$ is only obtained if $Y_{1,i} = 0$. We conclude that, when $Y_{2,i}$ is continuous, $Y_{1,i}$ is fully observed and $Y_{2,i}$ is instead only observed when $Y_{1,i} = 1$. When, instead, $Y_{2,i}$ is discrete and $P(Y_{2,i} = 0) > 0$, $y_i = 0$ does not imply that $Y_{i,1} = 0$. In this scenario a zero outcome can be either a chance finding of the counting process, i.e. $Y_{1,i} = 1$ and $Y_{2,i} = 0$, or the natural consequence of the sampled community being disease free, i.e. $Y_{1,i} = 0$. Although it can be empirically difficult to distinguish between these two phenomena under this scenario, unless there is a good scientific reason, we would not recommend the use of a discrete model for $Y_{2,i}$ such that $P(Y_{2,i} = 0) = 0$ which would only conceal the issue.

In the applications of Section 8.4.1 and 8.4.2, we carried out maximization of the likelihood function in (8.4) using Monte Carlo maximum likelihood as described in Section 5.2.1.2 in the case of generalized linear geostatistical models.

8.3   Spatial prediction

Based on model (8.4), we can define three spatially continuous predictive targets.

If we are interested in understanding the spatial structure of the zero-inflation process, our interest is in the prediction of $\pi(x)$. A possible interpretation that we can attach to $\pi(x)$ is that this expresses how likely a location x is to have suitable environmental conditions for the transmission of the disease.

If, instead, our interest lies in the prediction of disease risk, given suitable conditions for transmission, we would then require predictive inferences on the properties of the counting process $Y_{2,i}$. For example, if $Y_{2,i}$ is Binomial, our inferential goal is to predict $p(x)$, which we interpret as the probability of testing positive for a disease, given that transmission of the disease occurs at x.

Finally, if we want to understand the overall process of transmission, our inferential target are the combined properties of the zero-inflation and counting processes. In the case of prevalence data, we would then map the product $\pi(x)p(x)$.

However, depending on the public health problem to address, different pre-

dictive targets might be required in place of or in addition to those previously
outlined; see the application of Section 8.4.2 for an example.

8.4   Applications

8.4.1   River blindness mapping in Sudan and South Sudan

We analyse data on river-blindness nodule prevalence from Sudan and South
Sudan; these data were described in Section 1.1 for the whole of Africa. The
semi-arid Sahel region, which crosses the two countries, marks the transition
between the Sahara desert and the Sudanian savannah. The former is a hot
desert with permanent dissolution of clouds allowing unhindered light and
thermal radiation; the latter, instead, is covered in tropical forest, swamps,
and grassland, thus presenting more favourable environmental conditions for
completion of the life cycle of the *Onchocerca volvulus* parasite. Additionally,
the Nile river, which passes trough the Sahel in the Eastern part of Sudan,
and the White Nile, which flows from South Sudan into the Sudanese capital
Khartoum where it meets the Blue Nile, might also provide breeding sites,
created by floods, for the river-blindndess vector, a black fly of the *Simulium*
type.

Figure 8.2 shows a physical map of the area with the geographical locations
of the 900 sampled communities, labelled as green points if no case was found
among the examined individuals and as red points if at least one individual
had nodules. As expected, most of the locations with at least one reported
case for the disease are found in the southern part, below the Sahara desert,
while only a few locations with at least one case can be found in proximity of
the Nile river in the Northern region.

We fit two geostatistical models to the data: a standard Binomial geostatistical model with logistic link function; a zero-inflated Binomial model as in
(8.1) such that:

$$\log\left\{\frac{p(x_i)}{1 - p(x_i)}\right\} = \beta + S(x_i) + Z_i,$$

where Z_i is Gaussian noise with variance τ^2, $S(x)$ is a Gaussian process having
variance σ^2 and exponential correlation function with scale ϕ;

$$\log\left\{\frac{p(x_i)}{1 - p(x_i)}\right\} = \gamma + T(x_i),$$

where $T(x)$ is a Gaussian process, independent of $S(x)$, with variance ν^2 and
exponential correlation function with scale δ. It is important to notice that
the zero-inflated model is an extension of and, therefore, more flexible than
then the standard geostatistical model, which is recovered by setting $\pi(x) = 1$
for all x.

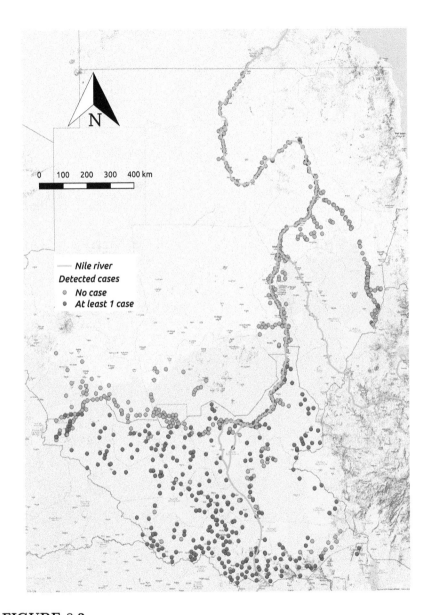

FIGURE 8.2
Map of the 900 sampled locations in Sudan and South Sudan. The background is a physical map of the region with the solid light blue line representing the Nile river.

TABLE 8.1

Monte Carlo maximum likelihood estimates and association 95% confidence intervals for the standard and zero-inflated geostatistical models of Section 8.4.1.

Parameter	Standard model		Zero-inflated model	
	Estimate	95% CI	Estimate	95% CI
β	-4.851	(-11.862, 2.160)	-1.819	(-2.618, -1.019)
$\log(\sigma^2)$	3.140	(1.413, 4.867)	-0.310	(-1.093, 0.474)
ϕ	7.392	(5.647, 9.136)	5.769	(4.911, 6.627)
$\log(\tau^2/\sigma^2)$	-3.295	(-5.051, -1.539)	-0.812	(-1.600, -0.025)
γ			-1.561	(-9.526, 6.404)
$\log(\nu^2)$			3.217	(1.337, 5.097)
$\log(\delta)$			7.565	(5.697, 9.434)

Table 8.1 shows the Monte Carlo maximum likelihood estimates from the two models. We observe that the points estimates and of σ^2 and ϕ and their associated 95% confidence intervals are largely comparable to the those for ν^2 and δ in the zero-inflated model. In the latter, instead, the estimates of σ^2, ϕ and τ^2/σ^2 are smaller in values and with narrower confidence intervals.

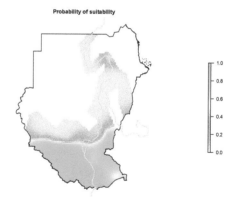

FIGURE 8.3

Map of the estimated probability of suitability for river-blindness, $\pi(x)$, from zero-inflated model of Section 8.4.1.

Figure 8.3 is a map of estimated probability of suitability, $\pi(x)$, for river-blindness. The map identifies two areas where $\pi(x)$ is close to zero in the north and in the south where $\pi(x)$ is close to one. However, there are also small pockets, one in the north-east, by the Nile river, and one in the south-east, where the estimated $\pi(x)$ is close to 0.5.

The differences in the resulting predictive inference for disease prevalence

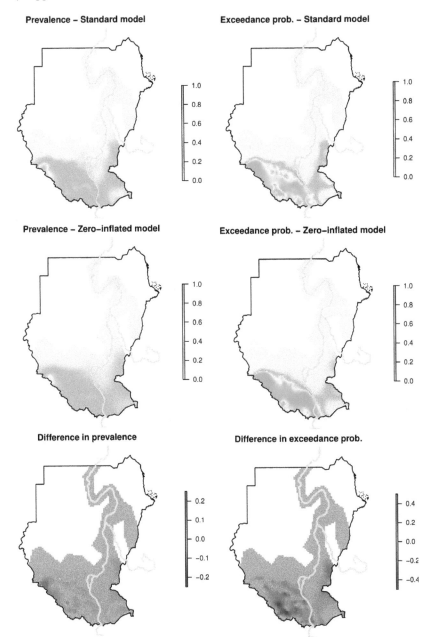

FIGURE 8.4

Surfaces of the estimated river-blindness nodule prevalence (left panels) and its probability of exceeding a threshold of 20% (right panels) from the standard (upper panels) and zero-inflated (middle panels) geostatistical models. The lower panels show the difference between the surfaces from the two models.

between the standard and the zero-inflated models are shown in Figure 8.4. (We remind that for the latter model that is given by the product $\pi(x)p(x)$.) Overall both models show qualitatively similar estimates of prevalence, with smaller values in the north and higher in the south. However, while for most of the prediction locations the point estimates are strongly similar (see left panels of Figure 8.4), for some the difference is large and ranges from around -0.2 (blue areas) to 0.25 (red areas). In the case of the exceedance probabilities (see right panels of Figure 8.4), the zero-inflated model shows less uncertainty that the standard model, with values closer to 1 in the south-west area.

From a statistical perspective, the zero-inflated model is an improvement over the standard model because its additional flexibility allows for more complex patterns in disease prevalence. The results of this analysis suggest that data show indeed a higher level of over-dispersion than that implied from the standard geostatsitical model for prevalence. The zero-inflated model allowed us to account for this by modelling the excess of zero cases as arising from a spatially structured process, thus providing more insights into environmental suitability for transmission of river-blindness across Sudan and South Sudan.

8.4.2 Loa loa: mapping prevalence and intensity of infection

Loiasis

- **The disease.** Loiasis, also known as the African eye worm, is a parasitic helminth disease caused by the nematode *Loa loa*. It is endemic in several countries across central and western Africa.

- **The vector.** It is transmitted from human to human through repeated bites of deerflies of the genus *Chrysops*. The deerflies are attracted by the movement of people and by smoke from wood fires. Rubber plantations may be an area where many deerflies are found.

- **The symptoms.** Most individuals infected with *Loa loa* present at most mild symptoms such as itchy Calabar swellings and the movement of adult worms across the eye.

- **The treatment.** Loiasis is usually treated with diethylcarbamazine, a medication also used in the treatment of other filarial diseases, including lymphatic filariasis and tropical pulmonary eosinophilia.

- **Source.** https://www.cdc.gov/parasites/loiasis

In this section we illustrate a re-analysis of Schlüter et al. (2016) concerning data on microfilarial loads of Loa loa (see text-box) in sampled individuals across villages of Cameroon, the Democratic Republic of Congo and the Republic of the Congo. The goal is to investigate the relationship between community-level prevalence and the proportion of highly infected individuals. The data in Schlüter et al. (2016) were collected in two field studies conducted in the West and East provinces of Cameroon (Takougang et al., 2002), and in the Republic of the Congo and the Bas-Congo and Orientale regions of the Democratic Republic of Congo (Wanji et al., 2012), respectively. In their analysis, the authors modelled the microfilariae (MF) counts per millilitre (ml) of blood from 19,128 individuals sampled across 222 villages. One of their objectives was to develop a statistical model to be used as an operational tool by public health workers, in order to predict the proportion of people in a village with an MF load exceeding a policy-relevant threshold of counts per ml of blood. An important feature of the data was the excess of zero counts, which invalidate the use of standard statistical models for count data.

Let $Y_j(x_i)$ denote MF density, measured as the number of MF per ml in a blood sample, for the j-th sampled individual at the village location x_i. In their non-spatial analysis, Schlüter et al. (2016) assume that $Y_j(x_i)$ is dependent on a latent bivariate zero-mean Gaussian random variable $Z(x_i) = (Z_1(x_i), Z_2(x_i))$, and that the $Z(x_i)$ at different locations are stochastically independent. Using the representation in (8.1), they then assume that conditionally on $Z_1(x_i)$ and $Z_2(x_i)$, $Y_{1,i}$ is a Bernoulli variable with probability of success $\pi(x_i)$, such that

$$\log\{\pi(x_i)/[1 - \pi(x_i)]\} = d(x_i)^\top \beta_1 + Z_1(x_i), \tag{8.5}$$

and $Y_{2,i}$ has a Weibull distribution with shape parameter κ and expectation $E[Y_2(x_i)] = \Gamma(1 + 1/\kappa)\mu(x_i)$, where $\mu(x_i)$ is modelled as

$$\log\{\mu(x_i)\} = d(x_i)^\top \beta_2 + Z_2(x_i). \tag{8.6}$$

This model has then been extended by Giorgi et al. (2017) as follows. Let $\mathcal{S}_h = \{S(x) : x \in A\}$, for $h = 1, 2$, and $\mathcal{T} = \{T(x) : x \in A\}$ denote a set of three independent stationary zero-mean Gaussian processes with unit variance. For each spatial process, we assume isotropic exponential covariance functions, hence

$$\text{corr}\{S_h(x), S_h(x')\} = \exp\{-u/\phi_{S_h}\}, h = 1, 2,$$

and

$$\text{corr}\{T(x), T(x')\} = \exp\{-u/\phi_T\},$$

where u is the Euclidean distance between x and x'.

We then assume that conditionally on $\mathcal{S}_1$, $\mathcal{S}_2$ and $\mathcal{T}$, the cumulative distribution function (cdf) of MF density $Y(x)$ at a village location x is given by

$$F\{y(x)\} = 1 - \pi(x) + \pi(x)G\{y(x); \kappa\}, \text{ if } y(x) > 0, \tag{8.7}$$

where $G\{\cdot; \kappa\}$ is a continuous cdf indexed by the parameter κ, and $\pi(x) =$

$1 - F(0)$ is the disease prevalence, at location x. We model $G\{\cdot; \kappa\}$ as a Weibull distribution with cdf

$$G\{y(x); \kappa\} = 1 - \exp\left\{-\left[\frac{y(x)}{\mu(x)}\right]^{\kappa}\right\},$$

where $\mu(x)$ is a spatially varying scale parameter and κ is a shape parameter, assumed to be common to all locations. Finally, we model the spatial variation in $\pi(x)$ and $\mu(x)$ as

$$\log\left\{\frac{\pi(x)}{1 - \pi(x)}\right\} = \mu_1 + \sigma_1[S_1(x) + T(x)]$$

and

$$\log\{\mu(x)\} = \mu_2 + \sigma_2[S_2(x) + T(x)],$$

where σ_1^2 and σ_2^2 are the variances of the linear predictors for prevalence and intensity, respectively. This constrains the component spatial processes to have the same variance whilst allowing the two linear predictors to have different variances. In this formulation, $T(x)$ contributes to the spatial variation of both prevalence and intensity whereas $S_1(x)$ and $S_2(x)$ account for independent residual spatial variation in prevalence and intensity, respectively.

The resulting standardized variogram for the linear predictors of prevalence and intensity is

$$\gamma_h(u) = 1 - \frac{1}{2}\left(\exp\{-u/\phi_{S_h}\} + \exp\{-u/\phi_T\}\right), \text{ for } h = 1, 2. \qquad (8.8)$$

Finally, using the definition of Cressie (1991, page 66, equation 2.3.19), the standardized cross-variogram between the two linear predictors is

$$\begin{aligned}
\gamma_{12}(u) &= \frac{1}{2}E[(S_1(x) + T(x) - S_2(x') - T(x'))^2] \\
&= 1 - \exp\{-u/\phi_T\}. \qquad (8.9)
\end{aligned}$$

TABLE 8.2
Monte Carlo maximum likelihood estimates and their 95% confidence intervals (CI) for the geostatistical model of Section 8.4.2.

	Estimate	95% CI
μ_1	-2.187	(-2.230, -2.144)
μ_2	8.258	(8.190, 8.327)
σ_1^2	0.874	(0.663, 1.152)
σ_2^2	0.146	(0.111, 0.193)
ϕ_S	17.982	(13.012, 24.850)
ϕ_T	154.520	(72.402, 329.774)
κ	0.552	(0.537, 0.568)

Table 8.2 reports MCML estimates for the model parameters. In fitting the model, we also set $\phi_{S_1} = \phi_{S_2} = \phi_S$, as a pragmatic strategy to circumvent a rather flat likelihood surface for these parameters. A Wald test of the null hypothesis that $\log\{\phi_{S_1}/\phi_{S_2}\} = 0$ give a non-significant result ($p > 0.05$).

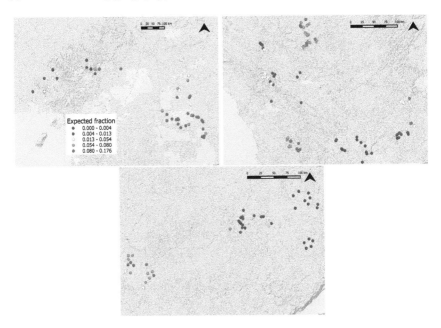

FIGURE 8.5
Expected fraction of individuals with more than 8000 MF counts per ml of blood, in each of the sampled viallges in the study sites in Cameroon (upper panel), the Republic of Congo (central panel) and the Democratic Republic of Congo (lower panel).

The estimates of μ_1, μ_2, σ_1^2, σ_2^2 and κ are comparable with those reported by Schlüter et al. (2016). Additionally, the processes S_1 and S_2 account for spatial variation in MF density up to about 35 km, beyond which their correlation function falls below 0.05, whilst the corresponding spatial range of the $\mathcal{T}$ process is about 300 km. This spatial structure is clearly visible from the map of point predictions (conditional expectations) of $R(x)$ shown in Figure 8.5.

Figure 8.6 shows the standardized empirical variograms and cross-covariograms based on the point predictions of $Z_1(x)$ and $Z_2(x)$ delivered by the non-spatial model of Schlüter et al. (2016). Each panel suggests the presence of residual spatial correlation, and does not show strong evidence against the fitted correlation structure.

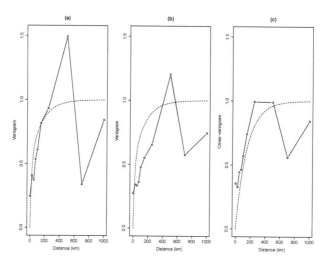

FIGURE 8.6

Empirical standardized variograms based on the predictive mean of the random effects associated with prevalence (a) and intensity (b), and their standardized cross-variogram (c), from the non-spatial model of Schlüter et al. (2016). The dashed lines represent the theoretical standardized variograms and cross-variogram from fitted geostatistical model to the Loa loa data, given by (8.8) and (8.9), respectively.

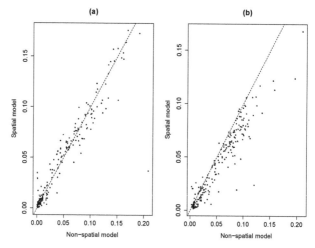

FIGURE 8.7

Scatter plot of the point estimates (a) and length of the 95% predictive intervals (b) for the prediction target, defined in (8.10) with $c = 8000$, from the non-spatial model of Schlüter et al. (2016) and the spatial model of Giorgi et al. (2017).

In this analysis, our predictive target, $R(x)$, is the probability that a randomly sampled individual at location x has an MF density above a predefined threshold c, hence

$$R(x_i) = \mathrm{P}\left\{Y(x_i) > c | W_i\right\} = \pi(x_i) \exp\left\{-\left[\frac{c}{\lambda(x_i)}\right]^{\kappa}\right\}, \quad \text{for } i = 1, \ldots, n.$$

$$(8.10)$$

The two panels of Figure 8.7 compare the point predictions and lengths of the 95% predictive intervals for the fraction of individuals with more than 8000 MF per ml of blood, from the fitted spatial and non-spatial models. Whilst the point estimates are in good agreement, the spatial model provides more accurate predictions, with lengths for the 95% predictive intervals almost always shorter than those from the non-spatial model. An intuitive explanation for this is that the spatial model can gain precision by borrowing strength of information from neighbouring villages.

9

Spatio-temporal geostatistical analysis

CONTENTS

In this chapter, we extend the geostatistical modelling framework illustrated in the previous chapters to the analysis of spatio-temporally referenced data. We shall adapt concepts and methods previously introduced to reflect the fact that *where* and *when* we observe a phenomenon of interest, both matters for the analysis of the data. Although the reader should bear in mind that spatio-temporal geostatistical analysis does not introduce any fundamentally new concept, unlike purely spatial analysis, it would be difficult - if not impossible - to formulate a model suitable for any spatio-temporal analysis. This is because context specific considerations into the possible interactions between space and time require different modelling approaches. We shall argue that selection between competing models should also be guided by the principle of *parsimony* whereby preference among models with the same explanatory power should be given to those with the least number of parameters.

In the remainder of this chapter, our focus will be on prevalence data from repeated cross-sectional surveys as these are one of the most commonly used study designs for disease surveillance, especially in low-resource settings where disease registries are non-existent or geographically incomplete. We emphasize that methods and principles, here illustrated, are not specific to prevalence data but holds for other data types.

9.1   Setting the context

We consider data obtained by sampling from a set of potential locations within an area of interest A, repeatedly at each of a sequence of times $t_1, \ldots, t_N$. At each sampled location, individuals are then tested for the disease under investigation. The data-format can be formally expressed as

$$\mathcal{D} = \{(x_{ij}, t_i, y_{ij}, n_{ij}) : x_{ij} \in A, j = 1, \ldots, m_i, i = 1, \ldots, N\}, \qquad (9.1)$$

where x_{ij} is the location of the jth of m_i sampling units at time t_i, n_{ij} is the number of tested individuals at x_{ij} and y_{ij} is the number of positively identified cases.

The methodology described in this paper can be equally applied to longitudinal or repeated cross-sectional designs. For this reason, we re-write (9.1) as

$$\mathcal{D} = \{(x_i, t_i, n_i, y_i) : x_i \in A, i = 1, \ldots, N^*\},$$

where $N^* = \sum_{i=1}^{N} m_i$ and either or both of the x_i and t_i may include replicated values.

An essential feature of the class of problems that we are addressing in this paper is that the locations x_i are a discrete set of sampled points within a spatially continuous region of interest. Another possible format for prevalence data, which we do not consider in the present study, is a small-area data-set. In this case, locations x_i are reference locations associated with a partition of A into n sub-regions. Disease registries in relatively well developed countries often use this format, both for administrative convenience and, in associated publications such as health atlases, to preserve individual confidentiality; see, for example, López-Abente et al. (2006) or Hansell et al. (2014).

A geostatistical model for data of the kind specified by (9.1) is that, conditionally on a spatio-temporal process $S(x, t)$ and unstructured random effects $Z(x, t)$, the outcomes Y are mutually independent binomial distributions with number of trials n and probability of being a case $p(x, t)$. Using the conventional choice of a logistic link function, although other choices are also available, we can then write

$$\log\left\{\frac{p(x_i, t_i)}{1 - p(x_i, t_i)}\right\} = d(x_i, t_i)^\top \beta + S(x_i, t_i) + Z(x_i, t_i), \qquad (9.2)$$

where $d(x_i, t_i)$ is a vector of spatio-temporally referenced explanatory variables with associated regression coefficients β. The spatio-temporal random effects $S(x_i, t_i)$ can be interpreted as the cumulative effect of unmeasured spatio-temporal risk factors. These are modelled as a Gaussian process with stationary variance σ^2 and correlation function

$$\mathrm{corr}\{S(x, t), S(x', t')\} = \rho(x, x', t, t'; \theta), \qquad (9.3)$$

where θ is a vector of parameters that regulate the scale of the spatial and temporal correlation, the strength of space-time interaction and the smoothness of the process $S(x,t)$. Finally, the unstructured random effects $Z(x_i, t_i)$ are assumed to be independent zero-mean Gaussian variables with variance τ^2, to account for extra-binomial variation within a sampling location. In particular applications, this can represent non-spatial random variation, such as genetic or behavioural variation between co-located individuals, spatial variation on a scale smaller than the minimum observed distance between sampled locations, or a combination of the two.

The model (9.2) can be used to address two related, but different, research questions.

Estimation: what are the risk factors associated with disease prevalence? In this case the focus of scientific interest is on the regression coefficients β.

Prediction: how to interpolate the spatio-temporal pattern of disease prevalence? The scientific focus is, in this case, on $d(x,t)^\top \beta + S(x,t)$ at both sampled and unsampled locations $\mathcal{X}$ and times $\mathcal{T}$. In some cases, the scientific interest may be more narrowly focused on $S(x,t)$, in order to identify areas of relatively low and high spatio-temporal variation that is not explained by the available explanatory variables.

Modelling of the residual spatio-temporal correlation through $S(x,t)$ is crucial in both cases: in the first case, in order to deliver valid inferences on the regression relationships by accurately quantifying the uncertainty in the estimate of β; in the second case, to borrow strength of information across observations y_i by exploiting their spatial and temporal correlation.

The use of explanatory variables $d(x,t)$ can also be beneficial in two ways: a simpler model for $S(x,t)$ can be formulated by explaining part of the spatio-temporal variation in prevalence through $d(x,t)$; more precise spatio-temporal predictions between data-locations also result from exploiting the association between disease prevalence and $d(x,t)$.

Our aim in this chapter is to provide a general framework that can be used as a tutorial guide to address some of the statistical issues common to any spatio-temporal analysis of data from prevalence surveys, especially when sampling is carried out over a large geographical area or time period, or both. More specifically, this chapter provides answers to each of the following research questions. How can we specify a parsimonious spatio-temporal model while taking account of the main features of the underlying process? How can we extend model (9.2) in order to account for non-stationary patterns of prevalence? What are the predictive targets that we can address using our model for disease prevalence? How can we effectively visualise the uncertainty in spatio-temporal prevalence estimates? These issues have only partly been addressed in current spatio-temporal applications of model-based geostatistics for disease prevalence mapping. Some of these are: Clements et al. (2006) on schistosomiasis in Tanzania; Gething et al. (2012) on the world-wide distribution of *Plasmodium vivax*; Hay et al. (2009) and Noor et al. (2014) on the world-wide and Africa-wide distributions of *Plasmodium falciparium*; R. W. Snow et al.

(2015) on historical mapping of malaria in the Kenyan Coast area; Bennett et al. (2013) on the mapping of malaria transmission intensity in Malawi; Kleinschmidt et al. (2001) on malaria incidence in KwaZulu Natal, South Africa; Kleinschmidt et al. (2007) on HIV in South Africa; Soares Magalhaes & Clements (2011) on anemia in preschool-aged children in West Africa; Raso et al. (2005) on schistosomiasis in Côte D'Ivoire; Pullan et al. (2011) on soil-transmitted infections in Kenya; Zouré et al. (2014) on river blindness in the 20 participating countries of the African Programme for Onchocerciasis control. In almost all of these cases, the adopted spatio-temporal model is only assessed with respect to its predictive performance, using ROC curves and prediction error summaries. In our view, a validation check on the adopted correlation structure in the analysis should precede geostatistical prediction, as misspecification of the spatio-temporal structure of the field $S(x,t)$ can potentially lead to an inaccurate quantification of uncertainty in the prevalence estimates and, therefore, to invalid inferences. In this paper, we describe the different stages of a spatio-temporal geostatistical analysis and provide tools that directly address the issue of specifying a spatio-temporal covariance structure that is compatible with the data.

9.2   Is the sampling design preferential?

Different design scenarios can give rise to data of the kind expressed by (9.1). A good choice of design depends both on the objectives of the study and on practical constraints.

In a longitudinal design, data are collected repeatedly over time from the same set of sampled locations. This is an appropriate strategy when temporal variation in the outcome of primary interest dominates spatial variation, and more obviously when the scientific goal is to understand change over time at a set of sentinel locations. A longitudinal design is also cost-effective when setting up a sampling location is expensive but subsequent data-collection is cheap.

In a repeated cross-sectional design, a different set of locations is chosen on each sampling occasion. This sacrifices direct information on changes in disease prevalence over time in favour of more complete spatial coverage. Repeated cross-sectional designs can also be adaptive, meaning that on any sampling occasion, the choice of sampling locations is informed by an analysis of the data collected on earlier occasions. Adaptive repeated cross-sectional designs are therefore particularly suitable for applications in which temporal variation either is dominated by spatial variation or can be well explained by available covariates; see Chapter 6.

To explain how the sampling design might affect our geostatistical analysis of the data, let $\mathcal{X} = \{x_i \in A : i = 1, \ldots, n\}$ denote the set of sampling locations arising from the sampling design, $\mathcal{S} = \{S(x) : x \in A\}$ the signal process and $\mathcal{Y} = \{Y_i : i = 1, \ldots, n\}$ the outcome data.

A sampling design is deterministic if it consists of a set of pre-defined sampling locations, and stochastic if the locations are a probability-based selection from a set of candidate designs. In the latter case $\mathcal{X}$ is a finite point process on the region of interest A. Let $[\cdot]$ denote "the distribution of." Our model for the outcome data is then obtained by integrating out $\mathcal{S}$ from the joint distribution $[\mathcal{X}, \mathcal{S}, \mathcal{Y}]$, i.e.

$$[\mathcal{X}, \mathcal{Y}] = \int [\mathcal{X}, \mathcal{S}, \mathcal{Y}] \, d\mathcal{S}. \tag{9.4}$$

From a modelling perspective, the most natural factorization of the integrand in the above equation is as

$$[\mathcal{X}, \mathcal{S}, \mathcal{Y}] = [\mathcal{S}][\mathcal{X}|\mathcal{S}][\mathcal{Y}|\mathcal{X}, \mathcal{S}]. \tag{9.5}$$

As explained in Chapter 7, the design is *non-preferential* if $[\mathcal{X}|\mathcal{S}] = [\mathcal{X}]$, in which case (9.4) becomes

$$[\mathcal{X}, \mathcal{Y}] = [\mathcal{X}] \int [\mathcal{S}][\mathcal{Y}|\mathcal{X}, \mathcal{S}] \, d\mathcal{S}. \tag{9.6}$$

Hence, under non-preferential sampling schemes, inference about $\mathcal{S}$ and/or $\mathcal{Y}$ can be conducted legitimately by simply conditioning on the observed set of locations, $\mathcal{X}$.

The simplest example of a probabilistic sampling design is completely random sampling. This can be interpreted, according to context, either as a random sample from a finite, pre-specified set of potential sampling locations or as an independent random sample from the continuous uniform distribution on A. Other examples include spatially stratified random sampling designs, which consist of a collection of completely random designs, one in each of a number of subdivisions of A, and systematic sampling designs, in which the sampled locations form a regular (typically rectangular) lattice to cover A, strictly with the first lattice-point chosen at random, although in practice this is often ignored.

Here as in other areas of statistics, the choice of sampling design affects inferential precision. If, for example, the inferential target is the underlying spatially continuous prevalence surface, $p(x, t^*)$ at a future time t^*, a possible design goal for geostatistical prediction would be to minimise the spatial average of the mean squared error,

$$\int_A \mathrm{E}[\{\hat{p}(x, t^*) - p(x, t^*)\}^2] dx,$$

where $\hat{p}(x, t^*)$ is a predictor for $p(x, t^*)$ obtained from (9.2). In contrast, a possible design goal for estimation of the relationship between a covariate $d(x, t)$

and disease prevalence would be to minimise the variance of the estimated regression parameter, $\hat{\beta}$.

Efficient sampling designs for spatial prediction generally require sampled locations to be distributed more evenly over A than would result from completely random or stratified random sampling; see, for example, Matérn (1986).

Stratified sampling often provides a more cost-effective design than simple random sampling from the general population. In cases where the strata correspond to sub-populations associated with different disease risk levels, a geostatistical model should account for the stratification through the use of an appropriate explanatory variable. To illustrate this, consider, for example, a population consisting of K strata which correspond to a partition of the region of interest, A, into non-overlapping regions $\mathcal{R}_k$ for $k = 1, \ldots, K$. We then take a random sample from each region $\mathcal{R}_k$ so that each location $x \in \mathcal{R}_k$ has probability of being selected proportional to the population of $\mathcal{R}_k$. If it is known that each of the strata $\mathcal{R}_k$ is associated with different levels in disease risk, this can be accounted for by including a factor variable in (9.2) with $K - 1$ levels or, if K is large, using random effects at stratum-level. In some cases the strata can also be grouped into sub-populations which are known to differ in their exposure to the disease. For example, let us assume that each stratum can be classified as being urban or rural and that these two types of areas are associated with different risk levels, i.e.

$$\log \left\{ \frac{p(x_i, t_i)}{1 - p(x_i, t_i)} \right\} = \beta + \alpha u(x_i) + S(x_i, t_i) + Z(x_i, t_i), \qquad (9.7)$$

where $u(x_i)$ is an indicator function that takes value 1 if $x_i \in \mathcal{R}_k$ and $\mathcal{R}_k$ is urban, and 0 otherwise. Under this model, it follows that

$$[\mathcal{Y}, \mathcal{S}, \mathcal{X}] = [\mathcal{X}][\mathcal{S}][\mathcal{Y}|\mathcal{S}, \mathcal{X}]$$

hence (9.7) does not constitute an instance of preferential sampling. This shows that variables used in the design should be included in the model when these are associated with the outcome of interest so as to ensure that the sampling is non-preferential.

Another common design in practice is the opportunistic sampling design (Hedt & Pagano, 2011), in which data are collected at convenient places, for example from presentations at health clinics, a market or a school. The limitations of this are obvious: opportunistic samples may not be representative of the target population and so not deliver unbiased estimates of $p(x, t)$. Also, as unmeasured factors relating to the disease in question are likely to affect an individudual's decision to present, the assumption of non-preferential sampling is questionable. For example, areas with atypically high or low levels of $p(x, t)$ may have been systematically oversampled; for a discussion and formal solution to the problem of geostatistical inference under preferential sampling, see Chapter 7.

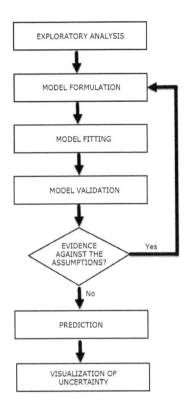

FIGURE 9.1
Diagram of the different stages of a statistical analysis.

9.3 Geostatistical methods for spatio-temporal analysis

In this section we describe a general framework for the analysis of data from spatio-temporally referenced prevalence surveys. Figure 9.1 shows the different stages of the analysis as a cycle that terminates when all the modelling assumptions are supported by the data. In our context, visualization of the results also plays an important role in order to display the spatio-temporal patterns of estimated prevalence and to communicate uncertainty effectively.

9.3.1   Exploratory analysis: the spatio-temporal variogram

As already shown in the case of a purely spatial geostatistical analysis, the
usual starting point for a spatio-temporal analysis of prevalence data is an
analysis based on a binomial mixed model without spatial random effects, i.e.
$S(x,t) = 0$ for all x and t. Let $\tilde{Z}(x_i, t_i)$ denote a point estimate, such as the
predictive mean or mode, of the unstructured random effects $Z(x_i, t_i)$ from
the non-spatial binomial mixed model. We then analyse $\tilde{Z}(x_i, t_i)$ to pursue
the two following objectives:

1.   testing for presence of residual spatio-temporal correlation;

2.   formulating a model for (9.3) and providing an initial guess for θ.

We make a working assumption that $S(x,t)$ is a stationary and isotropic
process, hence

$$\rho(x, x', t, t'; \theta) = \rho(u, v; \theta), \tag{9.8}$$

where $u = \|x - x'\|$, with $\|\cdot\|$ denoting the Euclidean distance, and $v = |t - t'|$.

The *variogram* can then be used to formulate and validate models for the
spatio-temporal correlation in (9.3). Let $W(x,t) = S(x,t) + Z(x,t)$, where
$S(x,t)$ and $Z(x,t)$ are specified as in (9.2); the spatio-temporal variogram of
this process is given by

$$\gamma(u, v; \theta) = \frac{1}{2} E[\{W(x,t) - W(x',t')\}^2] = \tau^2 + \sigma^2[1 - \rho(u, v; \theta)]. \tag{9.9}$$

We refer to this as the *theoretical* variogram, since it is directly derived
from the theoretical model for the process $W(x,t)$.

We use $\tilde{Z}(x_i, t_i)$ to estimate the unexplained extra-binomial variation in
prevalence, at observed locations x_i and times t_i. Let $n(u, v)$ denote the pairs
(i, j) such that $\|x_i - x_j\| = u$ and $|t_i - t_j| = v$; the *empirical* variogram is
then defined as

$$\tilde{\gamma}(u, v) = \frac{1}{2|n(u, v)|} \sum_{(i,j) \in n(u,v)} \{\tilde{Z}(x_i, t_i) - \tilde{Z}(x_j, t_j)\}^2, \tag{9.10}$$

where $|n(u, v)|$ is the number of pairs in the set.

Testing for the presence of residual spatio-temporal correlation can be
carried out using the following Monte-Carlo procedure:

(Step 1) permute the order of the data, including $\tilde{Z}(x_i, t_i)$, while holding
(x_i, t_i) fixed;

(Step 2) compute the empirical variogram for $\tilde{Z}(x_i, t_i)$;

(Step 3) repeat(i) and (ii) a large enough number of times, say B;

(Step 4) use the resulting B empirical variograms to generate 95% tolerance
intervals at each of the pre-defined distance bins.

If $\tilde{\gamma}(u, v)$ lies outside these intervals, then the data show evidence of residual spatio-temporal correlation. If this is the case, the next step is to specify a functional form for $\rho(u, v)$.

Gneiting (2002) proposed the following class of spatio-temporal correlation functions

$$\rho(u, v; \theta) = \frac{1}{(1 + v/\psi)^{\delta+1}} \exp\left\{-\frac{u/\phi}{(1 + v/\psi)^{\xi/2}}\right\}, \tag{9.11}$$

where ϕ and (δ, ψ) are positive parameters that determine the rate at which the spatial and temporal correlations decay, respectively. When $\xi = 0$ in (9.11), $\rho(u, v; \theta) = \rho_1(u)\rho_2(v)$ where $\rho_1(\cdot)$ and $\rho_2(\cdot)$ are purely spatial and purely temporal correlation functions, respectively. Any spatio-temporal correlation function that factorises in this way is called *separable*. In this sense, the parameter $\xi \in [0, 1]$ represents the extent of non-separability. M. L. Stein (2005) provides a detailed analysis of the properties of space-time covariance functions and highlights the limitations of using separable families. However, fitting of complex space-time covariance models requires more data than, in our experience, is typically available in prevalence mapping applications. In the application of Section 9.4, we show that only ψ and ϕ in (9.11) can be estimated with an acceptable level of precision, whilst the data are poorly informative with respect to the other covariance parameters, in which case the parsimony principle favours a separable model. Note, incidentally, that separability is implied by, but does not imply, that $S(x, t)$ can be factorised as $S_1(x)S_2(t)$, which would be a highly artificial construction.

A spatio-temporal correlation function is separable if

$$\rho(u, v; \theta) = \rho_1(u; \theta_1)\rho_2(v; \theta_2),$$

where θ_1 and θ_2 parametrise the purely spatial and temporal correlation functions, respectively; in the case of (9.11), this is separable when $\xi = 0$. Separable correlation functions are computationally convenient when joint predictions of prevalence are required at different time points over the same set of prediction locations. Checking the validity of the separability assumption can be carried out using the likelihood-ratio test for models such as (9.11), where separability can be recovered as a special case.

Once a parametric model has been specified, an initial guess for θ can be used to initialise the maximization of the likelihood function. One way to obtain an initial guess is to choose the value of θ that minimizes the sum of squared differences between the theoretical and empirical variogram ordinates. Section 5.3 of Diggle & Ribeiro (2007) describes the least squares algorithm and other, more refined methods to fit a parametric variogram model to an empirical variogram. However, in our view, variogram-based techniques should only be used for exploratory analysis and diagnostic checking. For parameter estimation and formal inference, likelihood-based and Bayesian methods are more efficient and more objective.

9.3.2   Diagnostics and novel extensions

In order to check the validity of the chosen spatio-temporal covariance function, we modify the Monte Carlo algorithm introduced in Section 9.3.1 by replacing (Step 1) with following.

(Step 1) Simulate $W(x_i, t_i)$ at observed locations x_i and times t_i, for $i = 1, \ldots, n$, from its marginal multivariate distribution under the assumed model. Conditionally on the simulated values of $W(x_i, t_i)$, simulate binomial data y_i from (9.2). Finally, compute the point estimates $\tilde{Z}(x_i, t_i)$ using the simulated data.

In this case, the resulting 95% tolerance band is generated under the assumption that the true covariance function for $S(x, t)$ exactly corresponds to the one adopted for the analysis. If $\tilde{\gamma}(u, v)$ lies outside the intervals, then this indicates that the fitted covariance function is not compatible with the data. To formally test this hypothesis, we can also use the following test statistic

$$T = \sum_{k=1}^{K} |n(u_k, t_k)| [\tilde{\gamma}(u_k, v_k) - \gamma(u_k, v_k; \theta)]^2, \tag{9.12}$$

where u_k and v_k are the distance and time separations of the variograms bins, respectively, the $n(u_k, t_k)$ are the numbers of pairs of observations contributing to each bin and θ is the true parameter value of the covariance parameters. Since θ is almost always unknown, it can be estimated using either maximum likelihood or Bayesian methods, in which case (9.12) should be averaged over the posterior distribution of θ using posterior samples $\theta_{(h)}$, i.e.

$$T = \frac{1}{B} \sum_{h=1}^{B} \sum_{k=1}^{K} |n(u_k, t_k)| [\tilde{\gamma}(u_k, v_k) - \gamma(u_k, v_k; \theta_{(h)})]^2. \tag{9.13}$$

The null distribution of T can be obtained using the simulated values for $\tilde{Z}(x_i, t_i)$ from the modified (Step 1) introduced in this section. Let $T_{(h)}$ denote the h-th sample from the null distribution of T, for $h = 1, \ldots, B$. Since evidence against the adopted covariance model arises from large values of T, an approximate p-value can be computed as

$$\frac{1}{B} \sum_{h=1}^{B} I[T_{(h)} > t],$$

where $I(a > b)$ takes value 1 if $a > b$ and 0 otherwise, and t is the value of the test statistic obtained from the data.

An unsatisfactory result from this diagnostic check could indicate a need for either or both of two extensions to the model: a more flexible family of stationary covariance structures; or non-stationarity induced by parameter variation over time, space or both.

In the former case, we note that the correlation function in (9.11) can also be obtained a special case of

$$\rho(u, v; \theta) = \frac{1}{(1 + v/\psi)^{\delta+1}} \mathcal{M}\left(\frac{u}{(1 + v/\psi)^{\xi/2}}; \phi, \kappa\right) \tag{9.14}$$

where $\mathcal{M}(\cdot; \phi, \kappa)$ is the Matérn (1986) correlation function with scale and smoothness parameters ϕ and κ, respectively (Gneiting, 2002). Equation (9.11) is recovered for $\kappa = 1/2$. However, the additional parameter introduced, κ, is likely to be poorly identified. A pragmatic response is to discretise the smoothness parameter κ in (9.14) to a finite set of values, e.g. $\{1/2, 3/2, 5/2\}$, over which the likelihood function is maximized.

In the second case, the context of the analysis can provide some insights on the nature of the non-stationary behaviour of the process being studied. For example, if data are sampled over a large geographical area, such as a continent, one may expect the properties of the process $S(x, t)$ to vary across countries. This can then be assessed by fitting the model separately for each country. A close inspection of the parameter estimates for θ might then reveal which of its components show the strongest variation. Furthermore, if these estimates also show spatial clustering, the vector θ, or some of its components, can be modelled as an additional spatial process, say $\Theta(x)$. The process $S(x, t)$ is then modelled as a stationary Gaussian process conditionally on $\Theta(x)$. A similar argument can also be developed if data are collected over a large time period in a geographically restricted area. In this case, θ may primarily vary across time and, therefore, could be modelled as a temporal stochastic process.

9.3.2.1   Example: a model for disease prevalence with temporally varying variance

We now give an example of how model (9.2) can be extended in order to allow the nature of the spatial variation in disease prevalence to change over time. We replace the spatio-temporal random effect $S(x, t)$ in the linear predictor with

$$S^*(x, t) = B(t)S(x, t), \tag{9.15}$$

where $B^2(t)$ represents the temporally varying variance of $S^*(x, t)$. We then model $\log\{B^2(t)\}$ as a stationary Gaussian process, independent of $S(x, t)$, with mean $-\eta^2/2$, variance η^2 and one-dimensional correlation function $\rho_B(\cdot; \theta_B)$, with covariance parameters θ_B. Note that, using this parametrisation, $E[B^2(t)] = 1$ and, therefore, $V[S^*(x, t)] = \sigma^2$. The resulting process $S^*(x, t)$ is a non-Gaussian process with heavier tails than $S(x, t)$ and correlation function

$$\text{corr}\{S^*(x, t), S^*(x', t')\} = \exp\{\eta^2(\rho_B(v; \theta_B) - 1)\}\rho(u, v; \theta). \tag{9.16}$$

The likelihood function is obtained as in (9.20) but now with $W(x_i, t_i) = S^*(x_i, t_i) + Z(x_i, t_i)$.

9.3.3 Defining targets for prediction

Let $\mathcal{P}(W^*) = \{p(x,t) : x \in A, t \in [T_1, T_2]\}$ denote the set of prevalence surfaces covering the region of interest A and spanning the time period $[T_1, T_2]$. Prediction of $\mathcal{P}$ is carried out by first simulating samples from the the predictive distribution of W^*, i.e. the distribution of W^* conditional on the data y. From each simulated sample of W^*, we then calculate any required summary, $\mathcal{T}$ say, of the corresponding $\mathcal{P}(W^*)$, for example means or selected quantiles at any (x,t) of interest. By construction, this generates a sample from the predictive distribution of $\mathcal{T}$. Computational details and explicit expressions can be found in Giorgi et al. (2017).

Two ways to display uncertainty in the estimates of prevalence are through quantile or exceedance probability surfaces. We define the *α-quantile surface* as

$$\mathcal{Q}_\alpha(W^*) = \{q(x,t) : P(p(x,t) < q(x,t)|y) = \alpha, x \in A, t \in [T_1, T_2]\}. \quad (9.17)$$

Similarly, we define the *exceedance probability surface* for a given threshold l as

$$\mathcal{R}_l(W^*) = \{r(x,t) = P(p(x,t) > l|y) : x \in A, t \in [T_1, T_2]\}. \quad (9.18)$$

Values of the point-wise exceedance probability $r(x,t)$ close to 1 identify locations for which prevalence is highly likely to exceed l, and vice-versa.

In public health applications, an exceedance probability surface is a suitable predictive summary when the objective is to identify areas that may need urgent intervention because they are likely to exceed a policy-relevant prevalence threshold, say l. A disease "hotspot" is then operationally defined as the set of locations x, at a given time t, such that $p(x, l) > l$.

In some cases, summaries by administrative areas can be operationally useful. For example, the district-wide average prevalence for a district D at time t is

$$p_t(D) = \frac{1}{|D|} \int_D p(x,t)\, dx, \quad (9.19)$$

where $|D|$ is its area of D. Incidentally, $p_t(D)$ can also be estimated more accurately than the point-wise prevalence $p(x,t)$, because it uses all the available information within D. Quantile and exceedance probability surfaces can be defined for $p_t(D)$ in the obvious way.

9.3.4 Accounting for parameter uncertainty using classical methods of inference

In this section, we review classical methods of inference and propose a way to incorporate parameter uncertainty into the spatio-temporal predictions in a non-Bayesian fashion.

Let $\lambda^\top = (\beta^\top, \sigma^2, \theta^\top)$ denote the set of unknown model parameters, including regression coefficients β, the variance σ^2 of $S(x,t)$ and covariance parameters θ. We use $[\cdot]$ as a shorthand notation for "the distribution

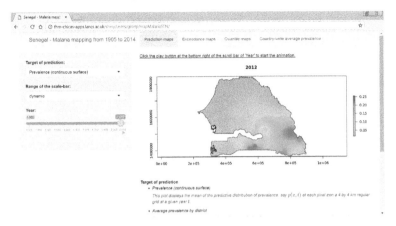

FIGURE 9.2
User interface of a Shiny application for visualization of results. The underlying data are described in Section 9.4.

of". The likelihood function is then obtained from the marginal distribution of the outcome $y^\top = (y_1, \ldots, y_n)$ by integrating out the random effects $W^\top = (W(x_1, t_1), \ldots, W(x_n, t_n))$ to give

$$L(\lambda) = [y|\lambda] = \int [W, y|\lambda] \, dW. \tag{9.20}$$

In general, the integral in (9.20) is intractable. However, numerical integration techniques or Monte Carlo methods can be used for approximate evaluation and maximization of the likelihood function; see Section 5.2 for more details.

Let W^* denote the vector of values of $W(x, t)$ at a set of unobserved times and locations. The formal solution to the prediction problem is to evaluate the conditional distribution of W^* given the data y. Although the joint predictive distribution of the elements of W^* is intractable, it is possible to simulate samples from this distribution.

If we assume, unrealistically, that λ is known, the predictive distribution of W^* is given by

$$
\begin{aligned}
[W^*|y, \lambda] &= \int [W^*, W|y, \lambda] \, dW \\
&= \int [W|y, \lambda][W^*|W, y, \lambda] \, dW \\
&= \int [W|y, \lambda][W^*|W, \lambda] \, dW. \tag{9.21}
\end{aligned}
$$

See Chapter 4 of Diggle & Ribeiro (2007) for explicit expressions.

If, more realistically, λ is unknown, *plug-in* prediction consists of replacing

λ in (9.21) by an estimate $\hat{\lambda}$, preferably the maximum likelihood estimate. A legitimate criticism of this is that the resulting predictive probabilities ignore the inherent uncertainty in $\hat{\lambda}$. However, this can be taken into account within a likelihood-based inferential framework as follows. Let $\hat{\Lambda}$ denote the maximum likelihood estimator of λ. We define the predictive distribution of W^* as

$$[W^*|y] = \int \int [\hat{\Lambda}][W|y, \hat{\Lambda}][W^*|W, \hat{\Lambda}] \, dW \, d\hat{\Lambda}, \qquad (9.22)$$

where $[\hat{\Lambda}]$ denotes the sampling distribution of the maximum likelihood estimator $\hat{\Lambda}$. Equation (9.22) acknowledges the uncertainty in $\hat{\Lambda}$ by expressing the predictive distribution $[W^*|y]$ as the expectation of the plug-in predictive distribution (9.21) with respect to the sampling distribution of $\hat{\Lambda}$. This can then be approximated using a multivariate Gaussian distribution with mean given by the observed MLE, $\hat{\lambda}$, and covariance matrix given by

$$\left[-\frac{\partial^2 \log L(\hat{\lambda})}{\partial^2 \lambda} \right]^{-1}.$$

In our experience, the quality of the Gaussian approximation is improved considerably by applying a log-transformation to each of the covariance parameters. If the Gaussian approximation remains questionable, a more computationally intensive alternative is a parametric bootstrap consisting of the following steps: simulate a number of binomial data-sets using the plug-in MLE for λ; for each simulated data-set, carry out parameter estimation by maximum likelihood. The resulting set of bootstrap estimates for λ can then be used to approximate the distribution of $\hat{\Lambda}$. We give an example of these approaches in the case-study of Section 9.4.

9.3.5   Visualization

The output from the prediction step consists of a set of N predictive surfaces, whether estimates, quantiles or exceedance probabilities, within the region of interest A at times $t_1 < t_2 < \ldots < t_N$. Animations then provide a useful tool for visualizing the predictive spatio-temporal surfaces and highlighting the main features of the interpolated pattern of prevalence. The R package `animation` (Xie, 2013) provides utilities for writing animations in several video and image formats. However, if interactivity is also desired, web-based "Shiny" applications (SAs) (RStudio, Inc, 2013) represent one of the best alternatives within R.

For the analysis carried out in Section 9.4, we have developed an SA which can be viewed at

`http://fhm-chicas-apps.lancs.ac.uk/shiny/users/giorgi/mapMalariaSEN/`.

The user-interface of this SA is shown in Figure 9.2. Any of four panels can be chosen in order to display predictive maps of prevalence ("Prediction maps"), exceedance probabilities with user defined prevalence thresholds ("Exceedance maps"), quantile surfaces ("Quantile maps") and country-wide summaries ("Country-wide average prevalence"). In the first three panels, the user can choose which target of prediction to display from a list and select the year on a slide bar. The range of prevalence and exceedance probabilities used to define the colour scale can be set to the observed range across the whole time series ("fixed") or specific to each year ("dynamic"). The former option is convenient for comparisons between years, whilst the latter gives a more effective visualization of the spatial heterogeneity in the predictive target in a given year.

9.4 Historical mapping of malaria prevalence in Senegal from 1905 to 2014

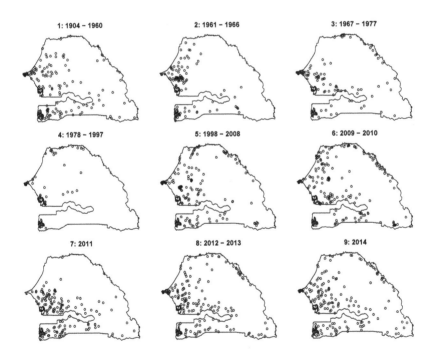

FIGURE 9.3
Locations of the sampled communities in each of the time-blocks indicated by Table 9.1.

We analyse malaria prevalence data from 1,334 surveys conducted in Senegal between 1905 and 2014. The data were assembled from three different data sources: historical archives and libraries of ex-colonial institutes; online electronic databases with data on malaria infection prevalence published since the 1980s; national household sample surveys. In assembling the data for the analysis, we only included locations that were classified as individual villages or communities or a collection of communities within a definable area that does not exceed 5 km^2. For more details on the data extraction, see R. Snow et al. (2015).

Malaria

- **The disease.** Malaria is a mosquito-borne disease caused by a parasite of the genus *Plasmodium*. It is widespread in tropical and subtropical regions of Asia, Latin America and sub-Saharan African, with the latter accounting for about 90% of cases and deaths globally.

- **The vector.** Malaria is most commonly transmitted by an infected female *Anopheles* mosquito. Such mosquitoes breed in a wide range of habitats but most species prefer clean, unpolluted water that usually found in salt-water marshes, mangrove swamps, rice fields, grassy ditches, the edges of streams and rivers, and small, temporary rain pools.

- **The symptoms.** Following between 8 and 25 days from infections, initial manifestations of malaria are similar to flu-like symptoms and viral disease, including headache, fever, shivering, joint pain, vomiting. One of the distinguishing symptoms of malaria is *paroxysm*, i.e. sudden coldness followed by shivering and then fever and sweating.

- **The treatment.** The recommended treatment for malaria is a combination of antimalarial medications that includes artemisinin, an antimalarial drug derived from the sweet wormwood plant, *Artemisia annua*.

- **Source.** www.cdc.gov/malaria/

The outcome of interest is the count y_i of positive microscopy tests out of n_i for *Plasmodium falciparum*, at a community location x_i and year t_i. Table 9.1 shows the number of surveys and the average prevalence for each of the indicated time-blocks. These were identified by grouping the data points so that each time-block contains at least 100 surveys. We observe that 649 out of the 1334 surveys were carried out between 2009 and 2014. Also, the empirical country-wide average prevalence steadily declines from the first to the last time-block. Figure 9.3 displays the sampled community locations within each of the time-blocks. The plot suggests a poor spatial coverage of Senegal in some years. The use of geostatistical methods can therefore be beneficial since it allows us to borrow the strength of information by exploiting the spatio-temporal correlation in the data.

TABLE 9.1
Number of surveys and country-wide average *Plasmodium falciparum* prevalence, in each time-block.

Time-block	Number of surveys	Average prevalence
1: 1904 - 1960	180	0.416
2: 1961 - 1966	109	0.384
3: 1967 - 1977	104	0.402
4: 1978 - 1997	101	0.134
5: 1998 - 2008	191	0.111
6: 2009 - 2010	187	0.051
7: 2011	140	0.043
8: 2012 - 2013	157	0.038
9: 2014	165	0.019

Our model for the data is of the form (9.23), with the following linear predictor

$$
\log\left\{\frac{p(x_i,t_i)}{1-p(x_i,t_i)}\right\} = \beta_1 + \beta_2 a(x_i,t_i) + \beta_3 \max\{a(x_i,t_i)-5,0\} +
$$
$$
\beta_4 A(x_i,t_i) + \beta_5 \max\{A(x_i,t_i)-5,0\} +
$$
$$
S(x_i,t_i) + Z(x_i,t_i), \tag{9.23}
$$

where $a(x_i,t_i)$ and $A(x_i,t_i)$ are the lowest and largest observed ages among the sampled individuals at location x_i and time t_i, respectively. In (9.23), we use *broken sticks*, each with a single knot, at 5 years for $a(x,t)$ and at 20 years for $A(x,t)$. For the spatio-temporal process $S(x,t)$, we use a Gneiting correlation function, as in (9.11), with $\delta = \xi = 0$, i.e. a separable covariance function.

Using the predictive mean as a point estimate of the random effects from a non-spatial binomial mixed model, we carry out the test for residual spatio-temporal correlation, as outlined in Section 9.3.1. The upper panels of Figure 9.4 show overwhelming evidence against the assumption of spatio-temporal independence. We then initialize the covariance parameters, ϕ and ψ, using a least squares fit to the empirical variogram, as shown by the dotted lines in the lower panels of Figure 9.4.

We conducted parameter estimation and spatial prediction using both likelihood-based and Bayesian inference. In the latter case, we specified the following set of independent and vague priors: $\beta \sim MVN(0,10^4 I)$; $\sigma^2 \sim \text{Uniform}(0,20)$; $\phi \sim \text{Uniform}(0,1000)$; $\tau^2/\sigma^2 \sim \text{Uniform}(0,20)$; $\psi \sim \text{Uniform}(0,20)$. Table 9.2 shows the maximum likelihood estimates of the

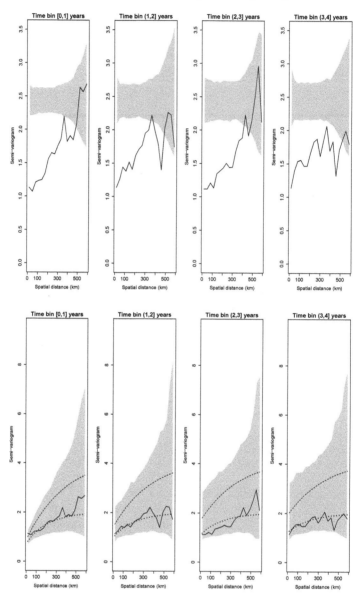

FIGURE 9.4
The plots show the results from the Monte Carlo methods used to test the
hypotheses of spatio-temporal indepence (upper panels) and of compatibility
of the adopted covariance model with the data (lower panels). The shaded
areas represent the 95% tolerance region under each of the two hypotheses.
The solid lines correspond to the empirical variogram for $\tilde{Z}(x_i, t_i)$, as defined in
Section 9.3.1. In the lower panels, the theoretical variograms obtained from the
least squares (dotted lines) and maximum likelihood (dashed lines) methods
are shown.

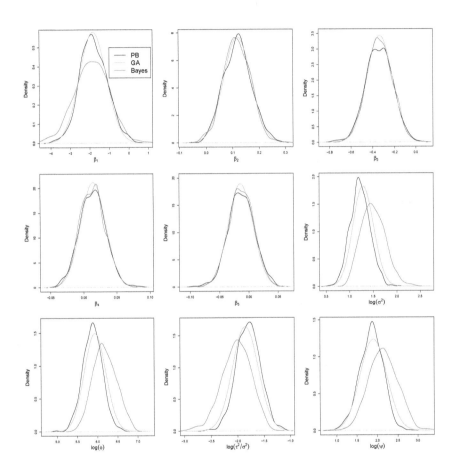

FIGURE 9.5
Density functions of the maximum likelihood estimator for each of the model parameters based on parameteric bootstrap (PB), as black lines, and the Gaussian approximation (GA), as orange lines; the blue lines correspond to the posterior density from the Bayesian fit.

TABLE 9.2
Maximum likelihood estimates of the model parameters and their 95% confidence intervals (CI) based on the asymptotic Gaussian approximation (GA) and parametric bootstrap (PB).

Parameter	Estimate	95% CI (GA)	95% CI (PB)
β_1	-1.830	(-3.180, -0.480)	(-3.131, -0.367)
β_2	0.118	(0.017, 0.220)	(0.019, 0.226)
β_3	-0.334	(-0.562, -0.105)	(-0.585, -0.103)
β_4	0.015	(-0.022, 0.052)	(-0.025, 0.052)
β_5	-0.014	(-0.055, 0.027)	(-0.056, 0.030)
σ^2	3.650	(2.378, 5.601)	(2.272, 5.222)
ϕ	381.022	(225.948, 642.528)	(220.593, 568.953)
τ^2/σ^2	0.157	(0.097, 0.253)	(0.105, 0.253)
ψ	6.730	(3.571, 12.683)	(3.484, 10.669)

TABLE 9.3
Posterior mean and 95% credible intervals of the model parameters from the Bayesian fit.

	Posterior mean	95% credible interval
β_1	-1.899	(-3.746, -0.275)
β_2	0.116	(0.013, 0.212)
β_3	-0.335	(-0.560, -0.115)
β_4	0.013	(-0.023, 0.050)
β_5	-0.013	(-0.054, 0.028)
σ^2	4.649	(2.887, 7.641)
ϕ	504.330	(283.019, 863.198)
τ^2/σ^2	0.137	(0.075, 0.217)
ψ	9.098	(4.443, 16.608)

model parameters and their corresponding 95% confidence intervals based on the Gaussian approximation (GA) and on parametric boostrap (PB), together with Bayesian esimates (posterior means) and 95% credible intervals. The two non-Bayesian methods give similar confidence intervals; the difference is noticeable, although still small in practical terms, only for the parameter ϕ. The Bayesian method gives materially larger estimates (see Table 9.3) of σ^2 and ϕ . Note that for both of these parameters, the prior means are substantially larger than the maximum likelihood estimates, suggesting that the priors, although vague, have nevertheless had some impact on the estimates.

Figure 9.5 gives a different perspective on the similarities and differences between the results obtained by the non-Bayesian and Bayesian methods. The Bayesian posterior density of the intercept has heavier tails than the sampling distribution of the maximimum likelihood estimator; the posterior densities of σ^2, ϕ and ψ are shifted to the right of their non-Bayesian counterparts, whilst the posterior density of τ^2/σ^2 is shifted to the left. Finally, there is some residual skewness in the PB distributions of the log-transformed covariance parameters.

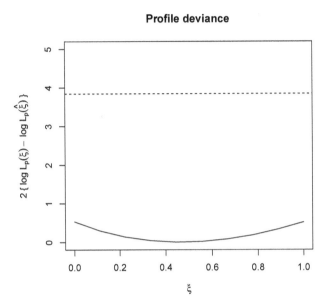

FIGURE 9.6
Profile deviance (solid line) for the parameter of spatio-temporal interaction ξ of the Gneiting (2002) family given by (9.11). The dashed line is the 0.95 quantile of a χ^2 distribution with one degree of freedom.

Using the Monte Carlo methods of Section 9.3.2, we checked the validity of the assumed covariance model. The lower panels of Figure 9.4 show that for each of the four time-lag intervals considered, the observed variograms fall within within the 95% tolerance region obtained under the fitted model; the p-value for a Monte Carlo goodness-of fit test using the test statistic (9.12) is 0.548.

Figure 9.6 shows the profile deviance function

$$D(\xi) = 2\{\log L_p(\hat{\xi}) - \log L_p(\xi)\},$$

where $L_p(\xi)$ is the profile likelihood for the parameter of spatio-temporal interaction parameter ξ and $\hat{\xi}$ is its Monte Carlo maximum likelihood estimate. The dashed horizontal line is the 0.95 quantile of a χ^2 distribution with one degree of freedom. The flatness of $D(\xi)$ indicates that data give very little information about the non-separability of the correlation structure of $S(x,t)$.

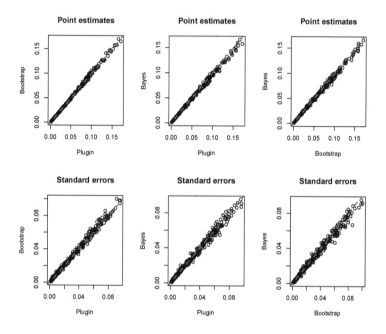

FIGURE 9.7
Scatter plots of the point estimates (upper panels) and standard errors (lower panels) of *Plasmodium falciparum* prevalence for children between 2 and 10 years of age, using plugin, parametric bootsptrap and Bayesian methods. The dashed red lines in each panel is the identity line.

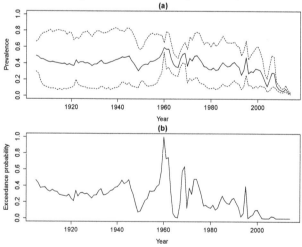

FIGURE 9.8
(a) Predictive mean (solid line) of the country-wide average prevalence with 95% predictive intervals. (b) Predictive probability of the country-wide average prevalence exceeding a 50% threshold.

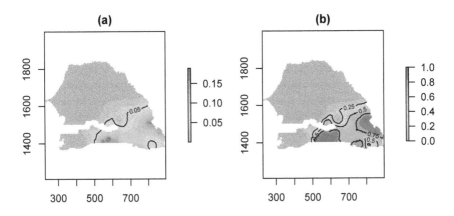

FIGURE 9.9
(a) Predictive mean surface of prevalence for children between 2 and 10 ($PfPR_{2-10}$); (b) Exceedance probability surface for a threshold of 5% $PfPR_{2-10}$. Both maps are for the year 2014. The contour lines correspond to 5% $PfPR_{2-10}$, in the left panel, and to 25%, 50% and 75% exceedance probability, in the right panel.

To assess the differences in the spatial predictions obtained using the GA, PB and Bayesian approaches, we used each method to predict *P. falciparum* prevalence for children between 2 and 10 years of age ($PfPR_{2-10}$) in the year 2014, at each point on a 10 by 10 km regular grid covering the whole of Senegal. Figure 9.7 shows pairwise scatterplots of the three sets of point predictions and associated standard deviations of $PfPR_{2-10}$. All six scatter plots show only small deviations from the identity line.

Figure 9.8(a) shows point and interval predictions of average country-wide $PfPR_{2-10}$. We observe a steady decline in $PfPR_{2-10}$ in the most recent decade. The highest predicted value of $PfPR_{2-10}$ across the whole of the time series occured in 1960, the year in which Senegal gained independence from France. Figure 9.8(b) shows for each year the predictive probability that average country-wide $PfPR_{2-10}$ exceeded 5%. Figure 9.9 shows the surfaces of the predictive mean (left panel) and the preditive probability that preva-lence exceeds 5% prevalence (right panel), for the year 2014. In the right panel, we can identify two disjoint areas in the south-west of Senegal, where the probability of exceeding 5% $PfPR_{2-10}$ is at least 75%. In areas between the contour of 50% and 75% exceedance probability we are less confident that $PfPR_{2-10}$ exceeds 5%. These aspects relating to the uncertainty about the 5% threshold cannot be deduced from the map of prevalence estimates in the left panel, nor would a map of pointwise prediction variances be of much help.

9.5   Discussion

We have developed a statistical framework for the analysis of spatio-
temporally referenced data from repeated cross-sectional prevalence surveys.
Our aim was to provide a set of tools and principles that can be used to identify a parsimonious geostatistical model that is compatible with the data. In
our view, model validation should include checking the validity of the specific
assumptions made on $S(x,t)$ rather than be focused exclusively on predictive performance, so as to avoid the risk of attaching spurious precision to
predictions from an inappropriate model.

The variogram is very widely used in geostatistical analysis. We use it
both for exploratory analysis and model validation, but favour likelihood-
based methods, whether non-Bayesian or Bayesian, for parameter estimation
and formal model comparison; an example of the latter is our use of the profile
deviance to justify fitting a model with separable correlation structure to the
Senegal malaria data.

In our spatio-temporal analysis of historical malaria prevalence data from
Senegal, we have shown how to incorporate parameter uncertainty within a
likelihood-based framework by approximation of the distribution of the maximum likelihood estimator using the Gaussian approximation and parametric bootstrap. The results showed that the Gaussian approximation provides
reliable numerical inferences for the regression coefficients but was slightly
inaccurate for the log-transformed covariance parameters. For this reason, we
generally recommend using parametric bootstrap whenever this is computationally feasible. In our view, this gives a viable approach to handling parameter uncertainty in predictive inference without requiring the specification
of so-called non-informative priors. Non-Bayesian and Bayesian approaches
showed some differences with respect to parameter estimation, but delivered
almost identical point predictions and predictive standard deviations for the
spatial estimates of prevalence. Our results also illustrate how even large geostatistical data-sets often lead to disappointingly imprecise inferences about
model parameters. For this reason, we woild favour Bayesian inference when,
and only when, an informative prior can be specified from contextually based
expert prior knowledge of the process under investigation.

In Section 9.3.2, we discussed how to extend the standard model for prevalence data in order to let the model parameters change over time, space or
both. However, the use of these models requires a large amount of the data
and good spatio-temporal coverage so as to detect non-stationary patterns in
prevalence. In the Senegal malaria application application the spatio-temporal
sparsity of the sampled locations meant that the data could not be used to
reliably detect spatio-temporal variation in the covariance parameters. For
this application we also assumed that the sampling locations did not arise
from a preferential sampling scheme. The standard geostatistical model for

prevalence can also be extended to account for preferentiality in the sampling design as shown in Chapter 7. However, such a model would require a larger amount of data than was available for this application.

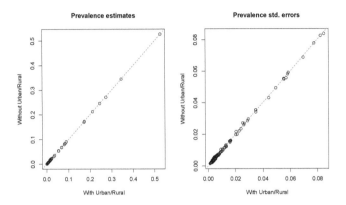

FIGURE 9.10
Prevalence estimates (left panel) and standard errors (right panel) based on the Demographic and Health Survey conducted in Senegal in 2014. Those are obtained from a model using a spatial indicator for urban and rural communities (x-axis) and excluding this explanatory variable (y-axis). The dashed line in both graphs is the identity line.

Our analysis included data from the Demographic and Health Survey (DHS) conducted in Senegal in 2014. These data were collected using a two-stage stratified sampling design (ANSD, 2015). In the first stage, 200 census districts (CDs) are randomly selected, 79 among urban CDs and 121 among rural CDs, with probability proportional to the population size. In the second stage, an enumeration list from each CD was used to sample households randomly. In the analysis reported above, we could not account for the sampling design of the DHS data because of the lack of information on urban and rural extents for every single year when the surveys were conducted. However, since this variable is available for 2014, we extracted the DHS data and fitted two geostatistical models with and without an explanatory variable that classifies every location as rural or urban. Figure 9.10 shows the plots for the estimated prevalence and associated standard errors obtained from the two models. The differences both in the point estimates and standard error of prevalence are negligible. Hence, we do not expect the sampling design adopted in the DHS survey to affect the results reported in Section 9.4.

In addition to the sampling designs that we discussed in Section 9.2, cluster sampling is another cost-effective alternative to simple random sampling. In households surveys, a cluster might correspond to a geographically restricted area, e.g. a village or group of households, which are randomly selected in a first stage. One of the potential, but still unexplored, uses of this sampling design in disease mapping would be to disentangle the long-range and small-range spatial variation in disease risk. To pursue this objective the nugget component $Z(x_i, t_i)$ in (9.2) could be modelled as an additional Gaussian process whose scale of spatial correlation is constrained to be smaller than that of $S(x_i, t_i)$. Separating these two spatial scales of correlation would require a large amount of data and would be dependent on the spatial arrangement of the clusters.

10

Further topics in model-based geostatistics

CONTENTS

10.1   Combining data from multiple surveys

In developing country settings, it is not always feasible to conduct randomised prevalence surveys, and when it is feasible their size and spatial coverage are often less than ideal because of resource constraints. This makes it attractive to consider supplementing the data from a randomised survey by data from one or more non-randomised surveys or "convenience samples", whereby data are gathered opportunistically. For example data may be collected at schools, markets or hospital clinics; a specific example would be to conduct tests for malaria parasitaemia in women presenting at ante-natal clinics.

Understanding the limitations of the sampling design adopted for each survey is clearly important in order to draw valid inferences from a joint analysis of the resulting data. In particular, non-randomized samples may reach an unrepresentative sub-population or be biased in other ways. Nonetheless, a combined analysis of data from randomised and convenience samples that estimates and adjusts for the bias in each convenience sample can be more efficient than an analysis that considers only the data from randomised surveys. This idea is obviously not specific to spatial problems; in a non-spatial context, Hedt & Pagano (2011) proposed a hybrid estimator of prevalence that supplements information from random samples with convenience samples, and showed that this leads to more accurate prevalence estimates than those available from using only the data from randomised surveys.

Giorgi et al. (2015) developed a multivariate generalized linear geostatistical model to account for data-quality variation amongst spatially referenced prevalence surveys. They assumed that at least one of the available surveys was known to deliver unbiased prevalence estimates. Analysis of the data from this survey might then reasonably be conducted using the standard model (5.6), which we here reproduce for convenience as

$$\log\left\{\frac{p(x_i)}{1-p(x_i)}\right\} = d(x_i)^\top \beta + S(x_i) + U_i. \tag{10.1}$$

Recall that the stochastic terms $S(x_i)$ and U_i in (10.1) represent spatial and non-spatial variation in prevalence that cannot be explained by the available covariates. Giorgi et al. (2015) then modelled the bias in a non-randomised survey using covariate information together with an additional, zero-mean stationary Gaussian process $\mathcal{B} = \{B(x) : x \in \mathbb{R}^2\}$. The resulting model for a non-randomised survey is

$$\log[p(x_j)/\{1 - p(x_j)\}] = d(x_j)'\beta + S(x_j) + U_j + \{d(x_j)'\delta + B(x_j)\}. \tag{10.2}$$

The indexing by j rather than i in (10.2) is intended to emphasise that multiple surveys typically collect data from different sets of locations. Data from both the randomised and the non-randomised survey then contribute to inference on the predictive target, $d(x)^\top \beta + S(x)$.

10.1.1 Using school and community surveys to estimate malaria prevalence in Nyanza province, Kenya

We now show an application to malaria prevalence data from a community survey and a school survey conducted in July 2010 in Rachuonyo South and Kisii Central Districts, Nyanza Province, Kenya (Stevenson et al., 2013). The analysis described here was previously reported in Diggle & Giorgi (2016).

The outcome variable in both surveys was presence/absence of malaria parasites by a rapid diagnostic test (RDT) using a finger-prick blood sample collected from each participant.

In the school survey, 46 out of 122 schools with at least 100 pupils were randomly selected using an iterative process to limit the probability of selecting schools with overlapping catchment areas. All children in attendance were eligible to be included.

In the community survey, residential compounds lying within 600 meters of each school were randomly sampled and all residents of each compound above the age of 6 months were eligible to be included.

The design of the school survey delivers an unbiased sample of schools, but children attending school may not be representative of the school's catchment area. In particular a plausible association between an individual's health status and their attendance at school would induce negative bias in school-based estimates of prevalence. The community survey, in contrast, delivers an unbiased sample of residents from the catchment area of each school. More details on the survey procedures can be found in Stevenson et al. (2013).

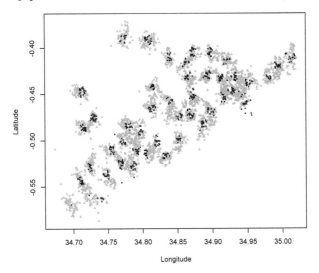

FIGURE 10.1
Geographical coordinates of the sampled compounds in the community (black points) and school (grey triangles) surveys.

In our analysis, we considered all sampled individuals between the ages of 6 and 25 years. This age range was represented in both surveys, as some adults have taken advantage of the introduction of free primary education in Kenya. Clearly, however, the resulting age-distribution of sampled individuals is not representative of the population at large. As malaria parasitemia is known to vary with age, adjustment for age is therefore essential to avoid unnecessary bias.

The community survey included 1430 individuals distributed over 740 compounds, whilst the school survey included 4852 pupils distributed over 3791 compounds. Figure 10.1 shows the locations of the sampled compounds from both surveys.

For our joint analysis of the data from the two surveys, we used exponential correlation functions for both $S(x)$ and $B(x)$, and denote by ϕ and ψ their respective scale parameters. We parameterise the respective variances of $S(x)$, $B(x)$ and U_i as σ^2, $\nu^2\sigma^2$ and $\omega^2\sigma^2$.

For selection of explanatory variables we used ordinary logistic regression, retaining variables with nominal p-values smaller than 5%. Table 10.1 gives

TABLE 10.1

Explanatory variables used in the analysis of malaria prevalence in Nyanza province, Kenya.

	Term
β_0	Intercept
β_1	Age in years
β_2	District (=1 if "Rachuonyo"; =0 otherwise)
β_3	Socio-economic status (score from 1 to 5)
δ_0	Survey indicator, 1 if "school," 0 if "community" (bias term)
δ_1	Age in years (bias term)

TABLE 10.2

Monte Carlo maximum likelihood estimates and corresponding 95% confidence intervals.

	Estimate	95% Confidence interval
β_0	-1.412	(-2.303, -0.521)
β_1	-0.141	(-0.174, -0.109)
β_2	2.006	(1.228, 2.785)
β_3	-0.121	(-0.169, -0.072)
δ_0	-0.761	(-1.354, -0.167)
δ_1	0.094	(0.046, 0.142)
$\log(\sigma^2)$	0.519	(0.048, 0.990)
$\log(\nu^2)$	-1.264	(-1.738, -0.790)
$\log(\phi)$	-3.574	(-4.083, -3.064)
$\log(\omega^2)$	-1.408	(-2.267, -0.550)
$\log(\psi)$	-3.366	(-4.178, -2.553)

the resulting set of explanatory variables. The "District" indicator variable accounts for a known higher level of malaria risk in Rachuonoyo district. Socio-economic status (SES) is an indicator of household wealth, taking discrete values from 1 (poor) to 5 (wealthy).

Table 10.2 reports Monte Carlo maximum likelihood estimates and 95% confidence intervals for the model parameters. The β-parameters reflect the district effect mentioned above as well as confirming a lower risk of malaria associated with higher scores of SES and greater age. The negative estimate of δ_0 and its associated confidence interval indicate, as anticipated, a significantly lower malaria prevalence in individuals attending school than in the community at large. The positive estimate and associated confidence interval

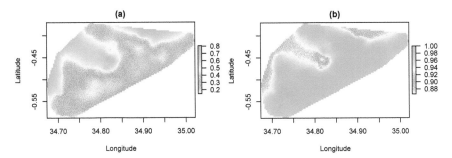

FIGURE 10.2
Maps of: (a) the point predictions of $B^*(x)$, the bias surface associated with the school-based sample of Kenyan children; (b) $r(x)$, the predictive probability that $B^*(x)$ lies outside the interval from 0.9 to 1.1 ().

for δ_1 indicate that the negative effect of age is less strong for individuals attending school than in the community.

Figure 10.2(a) shows point-wise predictions of the surface of $B^*(x) = \exp\{B(x)\}$, which represents the unexplained multiplicative spatial bias in the school survey for the odds of malaria at location x. Figure 10.2(b) maps the predictive probability that $B^*(x)$ lies outside the interval $(0.9, 1.1)$, i.e.

$$r(x) = 1 - P\left(0.9 < B^*(x) < 1.1 | y\right). \tag{10.3}$$

The lowest value of $r(x)$ is about 87%, indicating the presence of non-negligible spatially structured bias throughout the study area. The joint analysis of the data from both surveys allows us to adjust for the bias and so obtain more precise predictions for $S(x)$ than would be obtained using only the data from the community survey. To illustrate this, Figure 10.3(a) shows a scatter plot of the standard errors for $S(x)$ obtained from the joint model for the school and community surveys and from the model fitted to the community data only. Figure 10.3(b) partitions the study-region into areas where predictive standard errors from the joint analysis were smaller (green) or larger (red) than those from the analysis of the community survey data alone. Comparison with Figure 10.1 shows that the red areas correspond to unsampled parts of the study-region.

Giorgi et al. (2015) extended the above framework to a situation in which a sequence of randomised surveys were to be combined with a single, potentially biased convenience sample taken co-temporaneously with one of the randomised surveys. This extension is shown schematically in Figure 10.4. Nodes labelled Y_1, Y_2 and Y_2^* represent observed prevalences from unbiased surveys at times t_1 and $t_2 > t_1$, and from a biased survey at time t_2, respectively. Nodes labelled S_1 and S_2 represent the underlying prevalences at times t_1 and t_2. The node labelled B_2 represents the bias term for a convenience

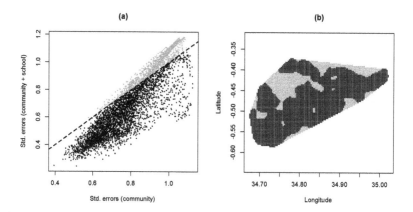

FIGURE 10.3
(a) scatter plot of the standard errors for $S(x)$ using models fitted to the data from the community survey against the community and school surveys, with dashed identity line; (b) plot of the prediction locations where black (grey) points correspond to locations where a reduction (an increase) in the standard errors for $S(x)$ is estimated when using the data from both the community and school surveys.

sample taken at time t_2. The targets for predictive inference are the prevalence surfaces S_1 and S_2.

10.2   Combining multiple instruments

It is often the case that prevalence data from a geographical region of interest are obtained using different diagnostic tests for the same disease under investigation. The reasons for this are manifold. For example, when the goal of geostatistical analysis is to map disease risk on a continental or global scale by combining data from multiple surveys, dealing with the use of different diagnostic tests may be unavoidable. In other cases, gold-standard diagnostic tests are often expensive and require advanced laboratory expertise and technology which may not always be available in constrained resource settings. This requires the use of more cost-effective alternatives for disease testing in order to attain a required sample size. Different diagnostics might also provide complementary information of intrinsic scientific interests into the spatial variation of disease risk and the distribution of hotspots.

To address the issue of combining prevalence data from multiple diag-

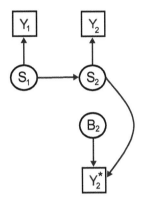

FIGURE 10.4
Representation of the Giorgi et al. (2015) model as a directed acyclic graph.

nostic techniques, Amoah et al. (2018) develop a geostatistical framework by distinguishing two main classes of inferential problems.

10.2.1 Case I: Predicting prevalence for a gold-standard diagnostic

A first class includes those problems where the goal is to carry prediction for a gold-standard diagnostics using a low-cost, but potentially biased alternative test.

An example of this is given for Loiasis, in the text-box "Diagnosis of two tropical diseases". In this case microscopy-based diagnosis for the disease represents the gold-standard technique. However, its use requires the availability of laboratories and trained technicians which are rarely available in remote, hard-to-reach areas of Africa. To tackle this issue, the World Health Organization has developed a more economically feasible questionnaire-based test called RAPLOA (`http://www.who.int/apoc/cdti/raploa/en/`). Amoah et al. (2018) propose a bivariate geostatistical model that allows to estimate the calibration relationship between two diagnostics so as to map microscopy-based prevalence in areas where the RAPLOA questionnaire is the only feasible option.

Let y_{i1} and y_{i2} be the counts of positively tested individual using RAPLOA ($i = 1$) and microscopy ($i = 2$) out of n_i individuals sampled at location x_i. Conditionally on two independent stationary and isotropic Gaussian processes, $S_1(x)$ and $S_2(x)$, and Gaussian noise, Z_{i1} and Z_{i2}, Amoah et al. (2018) assume that y_{i1} and y_{i2} are the realizations of two independent Binomial distribution with probability of a positive test p_{i1} and p_{i2}, respectively. Using link functions $f_1\{\cdot\}$ and $f_2\{\cdot\}$, with domain on the unit interval $[0, 1]$ and image on the real line, the model is then expressed by

$$\begin{cases} f_1\{p_1(x_i)\} = d^\top(x_i)\beta_1 + S_1(x_i) + Z_{i1} \\ f_2\{p_2(x_i)\} = d^\top(x_i)\beta_2 + S_2(x_i) + Z_{i2} + \alpha f_1\{p_1(x_i)\}. \end{cases} \tag{10.4}$$

Selection of suitable functions f_1 and f_2 can be carried out, for example, by exploring the association between the empirical prevalences of the two diagnostics in order to identify transformations that render their relationship approximately linear. Alternatively, subject matter knowledge could be used to constrain the admissible forms for f_1 and f_2; see, for example, Irvine et al. (2016) who derive a functional relationship between MF and an immuno-chromatographic test for prevalence of lymphatic filariasis by making explicit assumptions on the distribution of worms and their reproductive rate in the general population.

Diagnosis of two tropical diseases

Malaria

- Rapid diagnostic tests (RDTs) detects specific antigens (proteins) produced by malaria parasites in the blood of infected individual. They are particularly useful in remote areas with limited access to good quality microscopy services.

- Polymerase chain reaction (PCR) is a laboratory technique that is used to make many copies of a specific DNA region. PCR allows to identify the presence of malaria parasites in blood samples with a high sensitivity and specificity.

Loiasis

- Microscopy examination allows to identify the presence of the *Loa loa* microfilariae in blood samples, which should be collected in accordance with diurnal periodicity of the parasite. The *Loiasis* parasite can indeed be found in the peripheral blood circulation during the day but resides in the lungs during the night.

- RAPLOA is a questionnaire-based test on the history of visible worms moving in the lower part of the eye, to predict whether or not *Loiasis* is present at high levels in an individual.

10.2.2   Case II: Joint prediction of prevalence from two complementary diagnostics

Different diagnostic procedures might also provide complementary insights into the epidemiology of a disease under investigation. Unlike the problems of the previous section, in this case all diagnostics are considered gold-standard and the objective now lies in the development of a joint model that allows to exploit their cross-correlation in order to deliver more accurate predictive inferences on disease prevalence.

For example, two commonly used techniques for the detection of *Plasmodium falciparum* - the deadliest species of the malaria parasite - are rapid diagnostic tests (RDT) and polymerase chain reaction (PCR). PCR is highly sensitive and specific, but its use is constrained by high costs and the need for highly trained technicians. RDT is simpler to use, cost-effective and requires minimal training, but is less sensitive than PCR (Tangpukdee et al., 2009). Recent studies have reported that PCR and RDT can lead to the identification of different malaria hotspots, i.e. areas where disease risk is estimated to be unexpectedly high (Mogeni et al., 2017). In this context, mapping of both diagnostics is of epidemiological interest since their effective use is dependent on the level of malaria transmission, with PCR being the preferred testing option in low-transmission settings (Mogeni et al., 2017).

When both are of interest, Amoah et al. (2018) propose a joint geostatistical model that, using the same notation of the previous section, can be expressed as

$$f_k\{p_k(x_i)\} = d(x_i)^\top \beta_k + \nu_k\big[S_k(x_i) + T(x_i)\big] + Z_{ik}. \qquad (10.5)$$

The spatial processes $S_k(x)$ and $T(x)$ accounts for unmeasured risk factors that are specific to each and common to both diagnostics, respectively.

Given the relatively large number of parameters, fitting the model may require a pragmatic approach. In order to identify a parsimonious model for the data, we recommend an incremental modelling strategy, whereby a simpler model is used in a first analysis (e.g. by setting $S_k(x) = 0$ for all x) and more complexity is then added in response to an unsatisfactory validation check, as described below.

10.3 Incomplete data

10.3.1 Positional error

In this section, we consider the problem of carrying out geostatistical inference when the spatial locations X are recorded with error. More specifically, we focus our attention on *geomasking* which is obtained by adding a stochastic or deterministic displacement to the spatial coordinates X of a geostatistical data-set. Armstrong et al. (1999) advocated geomasking as an improvement on the standard practice of aggregating health records to preserve the confidentiality of information about individuals who might otherwise be identified by their exact spatial location.

Here, we consider stochastic perturbation methods, as these are the most commonly used in practice. Let Y_i denote the random variable associated with the outcome of interest, measured at locations x_i, for $i = 1, \ldots, n$. Let $S(x)$ be a stationary, isotropic Gaussian process with mean zero, variance σ^2 and Matérn (1960) correlation function $\rho(u_{ij}; \phi)$ where $u_{ij} = \|x_i - x_j\|$ is the Euclidean distance between any two locations x_i and x_j, $\phi > 0$ is a scale parameter. Also, let be Z_i a set of mutually independent $N(0, \tau^2)$ random variables. Let us consider the standard linear geostatistical model

$$Y_i = d(x_i)^\top \beta + S(x_i) + Z_i : i = 1, \ldots, n. \tag{10.6}$$

where, for any location x, $d(x)$ is a vector of explanatory variables with regression coefficients β.

The *variogram* for the outcome Y_i is defined as

$$
\begin{aligned}
V(u_{ij}) &= \frac{1}{2} \text{Var}\{Y_i - Y_j\} \\
&= \frac{1}{2} E[\{(Y_i - d(x_i)^\top \beta) - (Y_j - d(x_j)^\top \beta)\}^2] \tag{10.7}
\end{aligned}
$$

The Rice distribution

The random variable U follows a $Rice(\nu, \sigma)$ if its density function is

$$f(u; \nu, \sigma) = \frac{u}{\sigma^2} \exp\left(-\frac{u^2 + \nu^2}{2\sigma^2}\right) I_0\left(\frac{u\nu}{\sigma^2}\right),$$

with $I_k(\cdot)$ is the modified Bessel function of the first kind with order k.

The mean of U is

$$E[U] = \sigma \sqrt{\frac{\pi}{2}} L(\nu^2/2\sigma^2)$$

where

$$L(x) = e^{x/2}\left[(1 - x)I_0(x/2) - xI_1(x/2)\right];$$

the variance is

$$\text{Var}[U] = 2\sigma^2 + \nu^2 - \frac{\pi\sigma^2}{2} L^2(-\nu^2/2\sigma^2).$$

A Rice variable is also obtained as

$$U = \sqrt{X_1^2 + X_2^2},$$

where X_1 and X_2 are independent Gaussian variables both with variance σ^2, and mean $\nu \cos\theta$ and $\nu \sin\theta$, respectively.

When $S(x)$ is stationary and isotropic, (10.7) reduces to

$$V(u_{ij}) = \tau^2 + \sigma^2\{1 - \rho(u_{ij})\}.$$

In the presence of positional error, the true location is an unobserved random variable, which we denote by X_i^*. We observe the realised value of the displaced location,

$$X_i = X_i^* + W_i,\qquad(10.8)$$

where the W_i represent the positional error process. We assume that the W_i are mutually independent random variables whose bivariate density is symmetric about the origin with variance matrix $\delta^2 I$; we call δ^2 the *positional error variance*. In what follows we will assume that W_i follows a Gaussian distribution, but the results in the remainder of this section hold for any other symmetric distribution.

Let $U_{ij} = \|X_i - X_j\|$ and $V_{ij} = \{(Y_i - d(x_i)^\top\beta) - (Y_j - d(x_j)^\top\beta)\}^2/2$. It follows from (10.6) and (10.7) that U_{ij} and V_{ij} are conditionally independent given $U_{ij}^* = \|X_i^* - X_j^*\|$. Using the notation $[\cdot]$ to mean "the distribution of" it then follows that

$$[V_{ij} \mid U_{ij}] = \int_0^\infty [V_{ij} \mid U_{ij}^*][U_{ij}^* \mid u_{ij}]dU_{ij}^*.\qquad(10.9)$$

Also, $[V_{ij} \mid U_{ij}^*] = V_Y(U_{ij}^*)\chi_{(1)}^2$ and $[U_{ij}^* \mid u_{ij}]$ follows a so-called Rice (1944) distribution with scale parameter $\sqrt{2}\delta$ which we denote as $Rice(u_{ij}, \sqrt{2}\delta)$; for more information on this probability distribution see the text box. Taking the expectation of (10.9) with respect to $[Y_i, Y_j | u_{ij}]$ gives the theoretical variogram

$$V_Y(u_{ij}) = \tau^2 + \sigma^2\{1 - E[\rho(U_{ij}^*) \mid u_{ij}]\},\qquad(10.10)$$

where $E[\cdot]$ denotes expectation with respect to U_{ij}^*. As $\delta \to 0$, $V_Y(u_{ij})$ converges to the true variogram $V(U_{ij}^*)$ given by (10.7), whereas as $\delta \to \infty$ the spatial correlation structure of the data is destroyed and $V_Y(u_{ij}) \to \tau^2 + \sigma^2$.

In (10.10), the expectation on the right-hand side is not generally available in closed form. An exceptional case is the Gaussian correlation function, $\rho(u_{ij}) = \exp\{-(u_{ij}/\phi)^2\}$, which is the limiting case of the Matérn correlation function as $\kappa \to \infty$. In this case,

$$E[\rho(U_{ij}^*) \mid u_{ij}] = \frac{1}{1 + (2r)^2}\exp\left\{-\left(\frac{u_{ij}}{\phi\sqrt{1 + (2r)^2}}\right)^2\right\},\qquad(10.11)$$

where $r = \delta/\phi$. Hence, the magnitude of the bias in variogram estimation induced by geomasking depends on the ratio between the standard deviation of the positional error distribution and the range parameter of the correlation function of $S(x)$. Additionally, as $u_{ij} \to 0$ in (10.11), $E[\rho(U_{ij}^*) \mid u_{ij}] \to \{1 + (2r)^2\}^{-1} < 1$ and, as $r \to \infty$, $E[\rho(U_{ij}^*) \mid u_{ij}] \to 0$. We conclude that for variogram estimation, the main effect of ignoring geomasking is to introduce bias into the estimates of τ^2 and ϕ.

To derive the likelihood function for the linear geostatistical model with positional error, we use the following notation: $Y = (Y_1, \ldots, Y_n)$ is the collection of all the random variables associated with our outcome of interest; $S = (S(x_1), \ldots, S(x_n))$ is the vector of the spatial random effects at the observed locations x_i for $i = 1, \ldots, n$; $X = \{X_1, \ldots, X_n\}$ and $X^* = \{X_1^*, \ldots, X_n^*\}$ are the perturbed and the true locations, respectively. We then factorize their joint distribution as

$$
\begin{aligned}
[Y, S, X, X^*] &= [Y \mid S, X, X^*]\,[S, X, X^*] \\
&= [Y \mid S, X^*]\,[S \mid X, X^*]\,[X, X^*] \\
&= [Y \mid S, X^*]\,[S \mid X^*]\,[X^* \mid X]\,[X],
\end{aligned}
$$

where: $[Y \mid S, X^*] = \prod_{i=1}^n [Y_i \mid S(X_i^*)]$; $[Y_i \mid S(X_i^*)]$ is Gaussian distribution with mean $S(X_i^*)$ and variance τ^2; $[S \mid X^*]$ is multivariate Gaussian with mean zero and covariance matrix Σ such that $[\Sigma]_{ij} = \sigma^2 \rho(U_{ij}^*; \phi, \kappa)$; and $[X_i^* \mid X_i]$ is a bivariate Gaussian distribution with mean X_i and covariance matrix $\delta^2 I_2$. Also, note that in the above equation $[Y \mid S, X, X^*] = [Y \mid S, X^*]$ because of the conditional independence between Y and X given X^*.

The likelihood function for the unknown vector of parameters $\psi = (\beta, \sigma^2, \phi, \tau^2)$ is

$$
\begin{aligned}
L(\psi) &= [Y, X; \psi] \\
&= \int\!\!\int [Y, X, X^*, S; \psi]\, dS dX^* \\
&= \int\!\!\int [Y \mid X, X^*, S; \psi]\,[S, X, X^*; \psi]\, dS dX^* \\
&= \int\!\!\int [Y \mid X^*, S; \psi]\,[S \mid X^*; \psi]\,[X^* \mid X,]\,[X]\, dS dX^* \\
&\propto \int\!\!\int [Y \mid X^*, S; \psi]\,[S \mid X^*; \psi]\,[X^* \mid X]\, dS dX^*, \quad (10.12)
\end{aligned}
$$

After integrating out S in (10.12), the final expression for the likelihood is

$$
L(\psi) \propto \int [Y \mid X^*]\,[X^* \mid X]\, dX^*, \quad\quad (10.13)
$$

where $[Y \mid X^*, \psi]$ is a multivariate Gaussian distribution with mean $D^* \beta$ and covariance matrix $\Sigma + \tau^2 I_n$; here, D^* denotes the matrix of covariates at the true locations X^*.

Fanshawe & Diggle (2011) propose to approximate (10.13) by Monte Carlo integration. Given ψ and δ, they draw B independent samples from $[X^* \mid X; \psi]$. The resulting approximation to the likelihood is then obtained as

$$
L(\psi) \approx \frac{1}{B} \left[Y \mid X_{(b)}^*; \psi \right]
$$

where $X^*_{(b)}$ is the b-th samples from $[X^* \mid X; \psi]$. Maximization of the above expression is computationally intensive since a single evaluation of the approximated likelihood has a computational burden of order $O\left(B \times n^3\right)$. For this reason, Fanshawe and Diggle conclude that reliable computation of the standard errors for the maximum likelihood estimates is infeasible.

To overcome these computational issues, Fronterrè et al. (2018) have proposed to approximate the likelihood in (10.13) using the composite likelihood method (Varin et al., 2011). This approach has been applied to standard geostatistical models to make computations faster when the number of spatial locations is demanding (e.g. Vecchia (1988), M. L. Stein et al. (2004) and Bevilacqua & Gaetan (2015)).

The resulting approximation is obtained by treating each of the pairs of bivariate densities as independent, to give

$$
\begin{aligned}
\log L(\psi) &\approx \log L_{CL}(\psi) \\
&= \sum_{i=1}^{n-1} \sum_{j=i+1}^{n} \log[Y_i, Y_j; \psi] \\
&= \sum_{i=1}^{n-1} \sum_{j=i+1}^{n} \log \int_0^\infty \left[Y_i, Y_j \mid U^*_{ij}\right] \left[U^*_{ij} \mid u_{ij}\right] dU^*_{ij}. \quad (10.14)
\end{aligned}
$$

Hence, computation of the approximate likelihood requires the integration of $n(n-1)/2$ univariate integrals.

Fronterrè et al. (2018) have also shown that the effects of ignoring geomasking on parameter estimation are stronger for larger values in the ratio $r = \delta/\phi$, leading to a larger overestimation of τ^2 and ϕ and underestimation σ^2. For this reason, geomasking procedures should always use the smallest acceptable value for r. High values of r weaken the structure of the spatial dependence in the data, thus leading to less accurate predictive inferences.

10.3.2 Missing locations

To formally define the problem of missing locations in a geostatistical context, let us consider the standard linear model as in (10.6). Now, write X^* for a set of unknown locations at which measurements Y^* have been made, and let $X = (\tilde{X}, X^*)$ and $Y = (\tilde{Y}, Y^*)$. The observed quantities are given in this case by $\tilde{X}$ and Y. For any set $\mathcal{X}$ of points $x \in \mathbb{R}^2$ write $S(\mathcal{X}) = \{S(x) : x \in \mathcal{X}\}$; hence, $S(X) = (S(\tilde{X}), S(X^*))$. Assume that X and $S = \{S(x) : x \in \mathbb{R}^2\}$ are stochastically independent; the joint distribution of X, Y and S is then

$$
\begin{aligned}
[X, Y, S] &= [S][X][Y|S, X] \\
&= [S][\tilde{X}][X^* | \tilde{X}][\tilde{Y} | S(\tilde{X})][Y^* | S(X^*)], \quad (10.15)
\end{aligned}
$$

Our assumption that the sampling design is non-preferential allows a straight-forward marginalisation of (10.15) to give

$$
\begin{aligned}
[X, Y] &= [\tilde{X}][X^*|\tilde{X}][\tilde{Y}|X][Y^*|Y, \tilde{X}] \\
&= [\tilde{X}][X^*|\tilde{X}][\tilde{Y}|\tilde{X}][Y^*|\tilde{Y}, X^*] \\
&= [\tilde{X}][X^*|\tilde{X}][Y|X]
\end{aligned}
\tag{10.16}
$$

where $[Y|X]$ is a multivariate Gaussian distribution with mean vector $\mu = (d(x_1)^\top \beta, \ldots, d(x_n)^\top \beta)$ and covariance matrix Σ with diagonal elements $\sigma^2 + \tau^2$ and off-diagonal elements $\sigma^2 \rho(u_{ij}; \phi)$, with u_{ij} denoting the Euclidean distance between any two locations x_i and x_j.

10.3.2.1   Modelling of the sampling design

Depending on the problem under investigation, the set of sampling locations might be the result of a natural process, for example the locations of nests in a colony of birds. Alternatively, they might be obtained by using a random or regular lattice designs, as it is often the case for household surveys or agricultural field trials, respectively. Knowledge of the underlying process generating the sampling locations should then be incorporated into the specification of the distribution $[\tilde{X}]$. We now briefly outline some approaches to this specification, and propose a non-parametric approach that can be used when information on the underlying sampling process is limited.

One approach would be to model $[\tilde{X}]$ as an inhomogeneous Poisson process over the region of interest $A \subseteq \mathbb{R}^2$, with intensity

$$
\lambda(x) = \exp\{d(x)^\top \beta\},
\tag{10.17}
$$

where $d(x)$ is a p-dimensional vector of spatial covariates, such as population density in the case of a randomised household survey, and β is the associated vector of regression coefficients.

When no information on the sampling design is available a non-informative uniform distribution could be used, hence $\lambda(x) = \lambda$ for all $x \in A$. An alternative approach is to estimate the intensity $\lambda(x)$ from the data, using a kernel method. Let x_1 and x_2 denote the coordinates of the horizontal and vertical axes for a given point $x \in \mathbb{R}^2$. Then, the kernel density estimate of $\pi(x)$, i.e. the marginal density function of any component of X, based on the observed locations X^* is given by

$$
\hat{\pi}(x) = \frac{1}{n} \sum_{i=1}^{n} K_H(x_1 - x_{1i}; x_2 - x_{2i})
$$

where $K_H(\cdot; \cdot)$ is a bivariate kernel with symmetric and positive definite 2 by 2 smoothing matrix H. If we choose a Gaussian kernel, then H is the variance matrix of a bivariate Gaussian density and

$$
K_H(x_1 - x_{1i}; x_2 - x_{2i}) = \frac{1}{2\pi|H|^{1/2}} \exp\left\{ -\frac{1}{2}(x - x_i)^\top H^{-1}(x - x_i) \right\}
$$

The elements of H can be estimated by optimising an estimate of the mean-square-error (Berman & Diggle, 1989). Alternatively, if we assume that X is an independent random sample from a bivariate Gaussian distribution, the optimal H in the sense of minimising the integrated mean-square-error is $H = n^{-1/6}V$ (Lucy et al., 2002), where V is the sample covariance matrix.

Giorgi & Diggle (2015) have shown that the use of a homogeneous Poisson process prior $[X^*]$ may result in a very diffuse predictive distribution with widespread regions of high density. The use of a kernel density estimate for $\pi(x^*|\tilde{x})$, based on the set of observed locations is useful when the empirical distribution of $\tilde{X}$ is spatially heterogeneous and the conditional distribution of the unknown locations X^* is expected to follow the same pattern. Conversely, an inhibitory process for $[X^*|\tilde{X}]$ is more appropriate when the context suggests that the complete set of locations $X = (\tilde{X}, X^*)$ is likely to show some degree of spatial regularity.

Appendices

A

Background statistical theory

CONTENTS

Throughout this book, we have emphasised the central place of the likelihood in principled statistical inference. Our aim in this Appendix is to summarise briefly the important aspects of statistical theory that underpin likelihood-based methods of inference. Recommended book-length accounts, aimed at epidemiological or statistical readers respectively, include Clayton & Hills (1993) or Pawitan (2001).

A.1 Probability distributions

In the physical sciences, the outcome of an experiment can sometimes be predicted exactly. If you measure how long it takes a stone to fall a vertical distance x metres under the influence of gravity, the laws of physics tell you the time taken will be $t = \sqrt{x/9.81}$ seconds. In the life sciences, things are rarely that simple. If you grow 10 plants of the same species under apparently identical conditions and measure their heights at maturity, you will get ten different answers. Whilst we can't predict exactly what heights our ten plants will reach, we can assign a probability distribution over the range of possible heights. In this situation height is a *random variable*.

Useful summaries of the properties of a random variable include its *expectation* and *variance*, conventionally denoted by μ and σ^2, respectively. These are the theoretical analogues of the sample mean and variance, $\bar{y} = (\sum_{i=1}^{n} y_i)/n$ and $s^2 = \sum_{i=1}^{n} \{(y_i - \overline{(y)})\}^2/(n-1)$, of a set of data-values $y_1, ..., y_n$.

Amongst the infinitely many probability distributions that we could construct, the following three are of particular interest in geostatistical modelling of health outcomes.

A.1.1   The Binomial distribution

A *binomial* random variable Y has probability distribution

$$p(y) = \binom{m}{y} \theta^y (1 - \theta)^{m-y} : y = 0, 1, ..., m. \qquad (A.1)$$

Each $p(y)$ is the probability that Y takes the corresponding value y, m is a positive integer and $0 < \theta < 1$. The binomial distribution has expectation $\mu = m\theta$ and variance $\sigma^2 = m\theta(1-\theta)$; see Figure A.1. Another way to express this is as a *variance-mean relationship*, $\sigma^2 = \mu(1 - \mu/m)$.

Our interest in the binomial distribution is that it is a natural model for prevalence data. In this context, m is the number of people tested for presence or absence of a particular condition, θ is the probability that any one of them will test positive and y is the number who do test positive.

A.1.2   The Poisson distribution

A *Poisson* random variable Y has probability distribution

$$p(y) = \exp(-\mu)\mu^y/y! : y = 0, 1, ... \qquad (A.2)$$

Again, each $p(y)$ is the probability that Y takes the corresponding value y, which now can be any non-negative integer. Consistent with earlier notation, the Poisson distribution has expectation μ, and its variance is also equal to μ.

The Poisson distribution, named after the French mathematician Baron Siméon Denis Poisson, is widely used to describe the variation in open-ended count data. It arises naturally in physics as a model for particle emissions in radioactive decay, but is also used as a convenient approximation to the Binomial distribution when m is large and θ is small by setting $\mu = m\theta$. See Figure A.2.

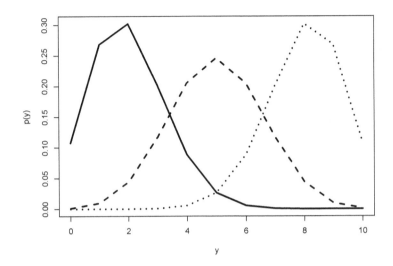

FIGURE A.1
Three examples of the binomial distribution, $p(y)$, with $m = 10$ and $\theta = 0.2, 0.5, 0.8$ (solid, dashed and dotted lines, respectively). The first and third are mirror images of each other. The second is symmetric about the point $y = 5$. In all three, the maximum value of $p(y)$ is at $y = m\theta$, the expectation of Y.

A.1.3   The Normal distribution

A *Normal* random variable Y has probability density[1]

$$f(y) = (2\pi\sigma^2)^{-\frac{1}{2}} \exp\left(-\frac{1}{2}(y - \mu)^2/\sigma^2\right) : -\infty < y < \infty. \qquad \text{(A.3)}$$

See Figure A.3. The expectation, μ, and variance, σ^2, of the Normal distribution are not functionally linked as was the case for the binomial and Poisson distributions. If Y has a Normal distribution, then so does $Y^* = a + bY$ for any specified real numbers a and b. Also, Y^* has expectation $a + b\mu$ and variance $b^2\sigma^2$. In particular, $Z = (Y - \mu)/\sigma$ has a *standard Normal distribution*, i.e. a Normal distribution with mean 0 and variance 1. A widely used shorthand notation for the Normal distribution is $N(0, \sigma^2)$.

The Normal distribution is also called the Gaussian distribution after the

---

[1] The change in terminology from "distribution" to "density" is because Y varies over the whole of the real line, and the probability of it taking any pre-specified exact value is zero. The probability that Y takes a value between any two pre-specified values a and b is obtained by integrating its probability density function, $f(y)$, hence $\text{Prob}(a < Y < b) = \int_a^b f(y)dy$.

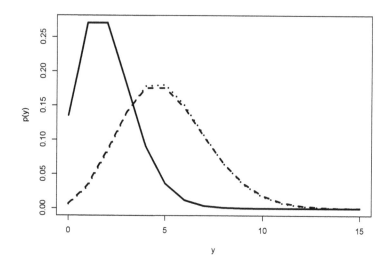

FIGURE A.2
Two examples of the Poisson distribution, $p(y)$, with $\mu = 2, 5$ (solid and dashed lines, respectively), and a binomial distribution with $m = 100$ and $p = 0.05$, hence expectation $\mu = 5$ (dotted line). The second and third are almost identical.

German mathematician C.F. Gauss (see Section 2.1). Its central place in statistical methodology is in part due to its mathematical tractability but also to the empirical fact that it often provides a good approximation to the behaviour of a real-valued random variable. Some theoretical support for this is provided by the remarkable *Central Limit Theorem*, which states roughly that averages of independent random variables are approximately Normally distributed, irrespective of their individual distributions. The precise statement is the following.

Central Limit Theorem. Let $Y_1, Y_2, ...Y_n$ be independent random variable, each with mean μ and variance σ^2, and define $Z_n = \sqrt{n}\bar{Y}/\sigma$ where $\bar{Y} = (\sum_{i=1}^{n} Y_i)/n$. Then, as $n \to \infty$ the distribution of Z_n tends to a standard Normal distribution.

A.1.4 Independent and dependent random variables

A set of random variables $Y_1, ..., Y_n$ are *independent* if the probability (or probability density) of any combination of values $y_1, ..., y_n$ is the product of their individual probabilities (or probability densities). This statement is the mathematical counterpart of the everyday use of the word, whereby the Y_i

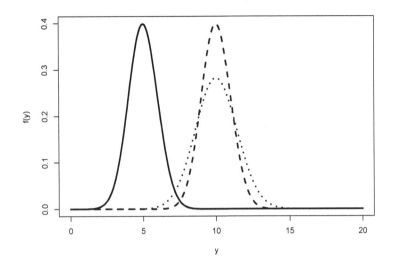

FIGURE A.3
The probability density functions, $f(y)$ of three Normal distributions, N(5, 1), N(10, 1), N(10, 4) (solid, dashed and dotted lines, respectively). Changing μ shifts $f(y)$ along the y-axis; changing σ^2 increases or decreases the spread of $f(y)$.

do not interfere with each other in any way. Random variables that are not independent are *dependent*.

Independent replication of an experiment or observation is a cornerstone of classical science, and of elementary statistical methods, but nature does not always act so obligingly: today's weather helps us to predict tomorrow's; finding that a disease has high prevalence in one community inclines us to think the same might be true in a neighbouring community.

In order to describe models for dependent random variables, we need to introduce the concepts of joint, marginal and conditional probability distributions or densities. To understand these ideas without unnecessary technical complication, we consider the special case of two random varialbes, and introduce a convenient shorthand notation, $[\cdot]$, to be read as "the distribution of." For example, $[Y] = \text{Poisson}(\mu)$ is read as "the distribution of the random variable Y is Poisson with expectation μ."

The separate distributions $[Y_1]$ and $[Y_2]$ are called the *marginal* distributions of Y_1 and Y_2. Now suppose that we first observe the value of Y_1, y_1 say, and are asked to predict the value of Y_2. If Y_1 and Y_2 are independent, knowing the value of Y_1 is of no help...all we have to help us is $[Y_2]$. In this setting, the *joint* distribution of Y_1 and Y_2 is $[Y_1, Y_2] = [Y_1][Y_2]$. If, on the other hand, Y_1

and Y_2 are dependent, knowing the one can help to predict the other, in the sense that observing the value of Y_1 *changes* the probability distribution of Y_2 from its marginal to its *conditional* distribution, which we write as $[Y_2|Y_1 = y_1]$ or just $[Y_2|Y_1]$. Notice that this is not one distribution but a whole set of potentially different distributions indexed by the possible values of Y_1. The joint distribution of Y_1 and Y_2 is now $[Y_1, Y_2] = [Y_1][Y_2|Y_1]$, and independence is the special case in which $[Y_2|Y_1] = [Y_2]$, for every possible value of Y_1.

What if we reverse the ordering of Y_1 and Y_2? The same argument gives the joint distribution of Y_1 and Y_2 as $[Y_1, Y_2] = [Y_2][Y_1|Y_2]$. It follows that if I choose to formulate a model for $[Y_1, Y_2]$ by specifying $[Y_1]$ and $[Y_2|Y_1]$, and you prefer to specify $[Y_2]$ and $[Y_1|Y_2]$ we have to obey certain rules if we want our respective models to be compatible with each other. Notice first that I can re-arrange your model as

$$[Y_1|Y_2] = [Y_1, Y_2]/[Y_2], \tag{A.4}$$

and that the probability of observing a particular value of Y_2 is the probability of observing that value in conjunction with *any* value of Y_1, hence $[Y_2] = \sum[Y_1, Y_2]$ where the summation is over all possible values of Y_1. Also, according to my model, $[Y_1, Y_2] = [Y_1][Y_2|Y_1]$. Plugging this into (A.4) gives

$$[Y_1|Y_2] = [Y_1][Y_2|Y_1]/\sum[Y_1][Y_2|Y_1]. \tag{A.5}$$

Equation (A.5), known as *Bayes' Theorem*, shows how our two models have to fit together if they are to agree with each other.

By far the most widely used family of joint distributions is the multivariate Normal or Gaussian distribution. A set of random variables $Y_1, ..., Y_n$ has a multivariate Normal distribution if any linear combination, $W = a_1Y_1 + ... + a_nY_n$ is Normally distributed, where the a_i are any specified set of real numbers. A very useful property of the multivariate Normal distribution is that its associated marginal and conditional distributions are also (univariate or multivariate) Normal. Specifically, the conditional distribution of any subset of the Y_i given the values y_j of the complementary sub-set is Normally distributed with expectations that are linear combinations of the y_j and a variance matrix that does not depend on the y_j.

A.2   Statistical models: responses, covariates, parameters and random effects

In very general terms, a statistical *model* is a specification of the joint probability distribution of a set of *output variables*, also called *response variables* and denoted by upper case Roman letters, typically $Y = (Y_1, ..., Y_n)$. In this book, each Y_i is a measurement associated with a location x_i in a designated geographical region. The form of the joint distribution typically depends on the values of one or more *input variables*, also called *explanatory variables* or *co-variates*, each denoted by a lower-case Roman letter, hence $d = (d_1, ..., d_n)$ etc, and on the values of one or more *parameters*, denoted by Greek letters $\alpha, \beta, ...$ Again in this book, each d_i is associated with a location x_i. and $d_i = d(x_i)$. These notational conventions are intended to emphasise that responses, covariate and parameters are very different things. A *response* is a random variable, i.e. the value that we measure cannot be predicted exactly in advance of its being taken. A *covariate* is a non-random variable. Its value is available in advance of its associated response being measured, either because it has been fixed in advance by the scientist or because it has been determined by some natural process that is *not* the process of scientific interest and, strictly, can be measured without error. A *parameter* is an unknown constant. Its value cannot be ascertained BY direct measurement (a literal translation of "para meter ($\pi\alpha\rho\alpha\mu\epsilon\tau\epsilon\rho$)" from Ancient Greek is "beyond measure"). Rather, it is a property of the scientific process of interest.

Example A.1. The simple logistic regression model
    To illustrate how a covariate, a response and one or more parameters fit together to define a statistical model, here is a simple (indeed, simplistic) statistical model for a hypothetical study of the relationship between the elevation (height above sea-level) of a location x and the prevalence of some disease of interest, $p(x)$ within a community at that location.
    At each of a set of locations $x_i : i = 1, ..., n$ take a sample of m_i people living at x_i, record the number Y_i who suffer from the disease in question and record the elevation, $d_i = d(x_i)$. Model each Y_i as a binomially distributed random variable with number of trials m_i and probability of "success" $p(x_i)$, and assume that

$$\log[p(x_i)/\{1 - p(x_i)\}] = \alpha + \beta \times d(x_i). \tag{A.6}$$

The two parameters of this model are α and β. Together, they describe how prevalence $p(x)$ varies with elevation, $d(x)$. For example, if β is positive, prevalence increases with elevation; if β is negative, prevalence decreases with elevation; if β is zero, prevalence is the same at all elevations and α is the log-odds of prevalence at any elevation.

    A fourth potential ingredient for a statistical model is a *random effect*. A random effect is a random variable whose value, unlike the value of a response, we cannot observe. An extension to Example A.1 illustrates the idea.

Example A.2. A logistic mixed effects model
    In the model (A.6), if our sample contains a number, k say, of communities whose elevation is the same, then the responses Y_i from these communities should behave as a random sample from a binomial distribution. Suppose

that we calculate the sample mean, $\bar{y}$ and the sample variance, s^2, of the observed responses from this set of communities. It follows from results stated in Section A.1.1 that we should observe $s^2 \approx \bar{y}(1 - \bar{y}/k)$. In practice, for real data of this kind it is not uncommon to observe that s^2 is materially bigger than $\bar{y}(1 - \bar{y}/k)$. A plausible explanation for this is that whilst these k villages share the same elevation, they differ in other ways. To capture this, we can modify (A.6) to

$$\log[p(x_i)/\{1 - p(x_i)\}] = \alpha + \beta \times d(x_i) + U_i, \qquad (A.7)$$

where the U_i are unobserved random variable, i.e. random effects. To complete the specification of the model, we need to choose a probability distribution for the U_i. The simplest choice is that they are independent, Normally distributed with mean zero (any non-zero mean being absorbed into α) and variance σ^2. In a spatial context, if we believe that the unobserved variable affecting prevalence has a geographical structure, a more realistic assumption might be that the U_i collectively follow a multivariate Normal distribution in which correlations between pairs of the U_i are determined by their geographical proximity.

The term *mixed effects model* refers to any model that includes both co-variates, also known as *fixed effects*, and random effects.

A.3   Statistical inference

A.3.1   The likelihood and log-likelihood functions

A probability distribution, such as the binomial or Poisson distributions defined in Section A.1, describes how, for any given values of the model's parameters, the probabilities of observing different responses vary. The corresponding *likelihood function* simply reverses the roles of parameter and response: describes how the probability of an observed response varies according to the values of the model's parameters. For example, the Poisson probability $p(y)$ defined at (A.2) can equally be written as a Poisson *likelihood*,

$$\ell(\mu) = \exp(-\mu)\mu^y/y! : \mu > 0. \qquad (A.8)$$

Unsurprisingly, the *log-likelihood* is defined to be the logarithm of the likelihood, hence

$$L(\mu) = -\mu + y\log(\mu) - \log(y!) : \mu > 0. \qquad (A.9)$$

Clearly, (A.8) and (A.9) contain equivalent information but we shall see that the log-likelihood is often the more convenient form to use

Suppose we observe a value $y = 10$ and we believe this to be an observation from a Poisson distribution with unknown mean μ. If μ were equal to 2, the

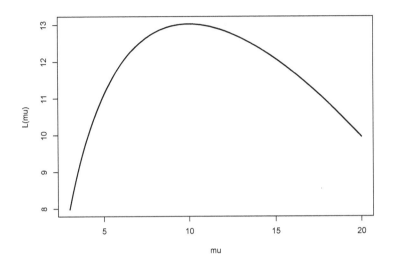

FIGURE A.4
The Poisson log-likelihood function for a single observation $y = 10$.

probability of observing $y = 10$ is 0.000038, i.e. highly unlikely. If, on the other hand, $\mu = 10$, the probability of observing $y = 10$ is 0.12511, i.e. rather more likely. Put another way, $\mu = 10$ is more compatible with the observation $y = 10$ than is $\mu = 2$. Figure A.4 plots the log-likelihood, which apart from a constant, $\log(10!)$, becomes

$$L(\mu) = -\mu + 10\log(\mu).$$

The log-likelihood takes its maximum value when $\mu = 10$; this is the value of μ that is the most compatible with the data.

Now suppose that we observe five values y_i that we believe to be a set of independent observations from a Poisson distribution with unknown mean μ, and that their sample mean is $\bar{y} = 10$. Independence of the observations implies that their joint probability is the product of their marginal probabilities, and the log-likelihood is therefore the sum of five terms of the form (A.9), one for each y_i, to give, again ignoring a constant term,

$$
\begin{aligned}
L(\mu) &= -5\mu + \log(\mu)\sum_{i=1}^{5} y_i \\
&= -5\mu + 50\log(\mu).
\end{aligned}
$$

This log-likelihood again takes its maximum value to $\mu = 10$, but is more concentrated than before around this value. To show this, Figure A.5 plots

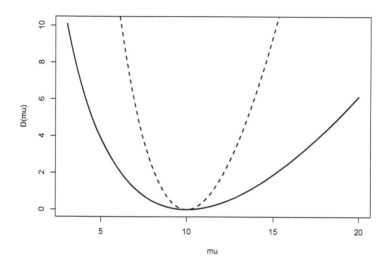

FIGURE A.5
The Poisson deviance function for a single observation $y = 10$ (solid line) and for a set of five independent observations with sample mean $\bar{y} = 10$ (dashed line).

the function $D(\mu) = 2\{L(10) - L(\mu)\}$, called the *deviance* function, for both cases. The deviance is simply the difference between the log-likelihood and its maximum possible value, scaled by a factor of two for theoretical reasons that we explain below.

The likelihood function is fundamental to both classical and Bayesian paradigms for statistical inference; where these two schools of inference differ is in how they interpret and use the likelihood function.

A.3.2   Estimation, testing and prediction

We have seen that a statistical model contains up to four different kinds of ingredients: responses, covariates, parameters and random effects. Amongst these, responses and covariates are known quantities – their values are included in a data-set. Parameters and random effects are unknown quantities. The fundamental goal of statistical inference is to use what is observed (data and model) to learn as much as possible about what is unknown (parameters and random effects). In the authors' view, a clear distinction between parameters and random effects are that parameters are properties of the underlying scientific process that generated the data, whereas random effects are properties of a particular realisation of that process, i.e. if an experiment is replicated under identical conditions the parameters of a statistical model for the experiment do not change, but the values of the random effects do. Within this view of statistical inference, learning about the unknown values of parameters is called *estimation*, whilst learning about the unknown values of random effects is called *prediction*.

Consider first estimation of a generic parameter, θ. A *point estimate* is a single best guess at the true value of θ. An *interval estimate* is a range of values that are broadly compatible with the data and assumed model; here, "compatibility" can be formally defined and quantified in two different ways, as we discuss in Sections A.3.3 and A.3.4.

A statistical *hypothesis test* can be thought of as a special case of interval estimation, in which a particular value of θ, say θ_0, has a special status in advance of data-collection, and the inferential goal is simply to establish the compatibility or otherwise of that particular value with the data and assumed model. The mechanism for doing this is to choose a *test statistic*, D say, which gives a numerical measure of the discrepancy between the data and the hypothesised value θ_0 and reject this hypothesised value if the probability of a value of D at least as big as its observed value when $\theta = \theta_0$ is smaller than a specified threshold value α; conventionally, $\alpha = 0.05$ corresponding to a so-called "5% level of significance."

For historical reasons, statistical tests feature prominently in many statistical courses and textbooks, especially those aimed at non-statisticians who need to analyse their own data. A limited justification for this is that in the pre-computer age, tests were sometimes the only convenient statistical tools available. Statisticians generally agree that estimation is able to address questions that are inherently richer, and often scientifically more relevant, than those that can be addressed with statistical tests and that the emphasis in statistical teaching should accordingly be much more on estimation and much less on testing. However, this is not to deny that a test can be a useful statistical tool for specific purposes; for a balanced discussion of this issue that remains relevant more than 40 years since its publication, see Cox (1977).

A.3.3 Classical inference

In classical inference, any parameter θ is considered as a constant whose value is unknown, but about which we can learn from data, y, by comparing different values of the log-likelihood function, $L(\theta)$. Values of θ that correspond to relatively large or small values of $L(\theta)$ are considered to be more or less supported by the evidence provided by the data, y. Thus, for a point estimate of θ we use the *maximum likelihood estimate*, $\hat{\theta}$, defined to be the value which maximises $L(\theta)$. Similarly, for an interval estimate of θ we use a *likelihood interval*, defined to be the set of values for which $L(\theta) \geq L(\hat{\theta}) - c$, for some suitable value of c.

Example A.3 The Poisson log-likelihood re-visited.

Suppose that random variables $Y_1, Y_2, ..., Y_n$ are independent and Poisson-distributed with parameter μ. Recall from Section A.1.2 that μ is the expectation of each Y_i and from Section A.1 that the expectation of a random variable is the analogue of the sample mean, $\bar{y}$. This might suggest that a reasonable guess at the unknown value of μ, having observed the data $y_1, y_2, ..., y_n$ might be the sample mean, i.e. $\hat{\mu} = \bar{y}$. What does the likelihood have to say? Since the Y_i are independent, the probability of observing a sequence $y_1, y_2, ..., y_n$ is the product of the n probabilities, hence the log-likelihood is the sum of the log-probabilities,

$$
\begin{aligned}
L(\mu) &= \sum_{i=1}^{n} \{-\mu + y_i \log(\mu)\} \\
&= -n\mu + (\sum_{i=1}^{n} y_i) \log(\mu) \\
&= -n\{\mu - \bar{y}\log(\mu)\}. \tag{A.10}
\end{aligned}
$$

Either by plotting $L(\theta)$ for different values of $\bar{y}$, or by differentiating $L(\theta)$ and solving the equation $L'(\mu) = 0$ it is easy to see that $L(\mu)$ is maximised when $\mu = \bar{y}$, i.e. the intuitive "guess" $\hat{\mu} = \bar{Y}$ is in fact the maximum likelihood estimate. Figure A.6 shows how we can calculate both the maximum likelihood estimate and a likelihood interval from a plot of the log-likelihood function (A.10). The constant c used to define the likelihood interval is $c = 1.92$, for reasons that will be explained shortly.

The log-likelihood is a function of the observed data y, and is therefore a realisation of a random variable whose distribution is induced by the distribution of Y. To emphasise this, we introduce an expanded notation $L(\theta, y)$ for the observed log-likelihood and $L(\theta, Y)$ for the corresponding random variable. Similarly, we write $\hat{\theta}(y)$ for an observed value of $\hat{\theta}$ and $\hat{\theta}(Y)$ for the corresponding random variable. The derivatives of the log-likelihood function with respect to elements of θ play an important role in classical inference. In particular, we define the *information matrix*, $I(\theta)$ to have (j, k)th element

$$
I_{jk} = E_Y \left[-\frac{\partial^2}{\partial\theta_j \partial\theta_k} L(\theta, Y) \right].
$$

The statistical properties of likelihood-based methods of inference are summarised by two important theorems.

Theorem A.1. $\hat{\theta}(Y) \sim \text{MVN}(\theta, I(\theta)^{-1})$

Theorem A.2. $D(\theta) = 2\{L(\hat{\theta}(Y), Y) - L(\theta, Y)\} \sim \chi_p^2$, where p is the dimensionality of θ.

Both theorems strictly apply only for large values of n, the number of data-

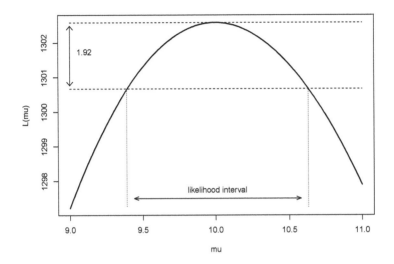

FIGURE A.6
Graphical calculation of a maximum likelihood estimate and a likelihood interval. The plotted function is the Poisson log-likelihood (A.10) for sample size $n = 100$ and sample mean $\bar{y} = 10$.

points, but can be used in practice under very general, albeit not universal, conditions. The principal circumstances in which the theorems do not apply are when the true value of θ is on a boundary of the parameter space (for example, a zero component of variance), when one or more elements of θ define the range of Y (for example, if the Y_i are uniformly distributed over the interval $(0, \theta)$) or when the dimensionality of θ is ambiguous. For detailed discussion, see for example Cox & Hinkley (1974).

An important practical point is that the accuracy of Theorem A.1 for finite data-sets can be improved by transforming the model parameters, typically to remove any restrictions on their permissible ranges. For example, in estimating a variance, the theorem usually gives a more accurate approximation by defining θ to be the logarithm of the variance; our **PrevMap** software does this routinely.

With the above caveats, Theorems A.1 and A.2 are used for interval estimation and testing as follows. For ease of explanation, we assume temporarily that θ is a single parameter, rather than a vector.

Consider first interval estimation. A *confidence interval* for a single parameter θ is a random interval constructed in such a way that, over repeated sampling of the data y from the underlying model, it will contain the true, fixed but unknown value of θ with a pre-specified probability p. Conventionally, confidence intervals are set using $p = 0.95$, called a "95% confidence interval."

Example A.4 Theorem A.1 and the Poisson log-likelihood.

Recall from Example A.3 that the log-likelihood for the parameter μ when the data $y = (y_1, ..., y_n)$ are independent, Poisson-distributed with parameter μ is

$$L(\mu) = -n\left\{\mu - \bar{y}\log(\mu)\right\},$$

and that the maximum likelihood estimator for μ is $\hat{\mu} = \bar{y}$. Differentiating $L(\mu)$ twice gives

$$\frac{d^2 L(\mu)}{d\mu^2} = -n\bar{y}/\mu^2.$$

According to Theorem A.1, $\hat{\mu}$ is Normally distributed with mean μ and variance $\mu^2/(n\bar{y})$, which we can approximate by $\hat{\mu}^2/(n\bar{y}) = \bar{y}/n$. It follows that with probability $p = 0.95$, $\hat{\mu}$ lies between $\mu - 1.96\bar{y}/n$ and $\mu + 1.96\bar{y}/n$. Equivalently, the interval $\hat{\mu} \pm 1.96\bar{y}/n$ contains μ with probability 0.95, i.e. it is a 95% confidence interval for μ. For example, with $n = 100$ and $\bar{y} = 10$ the 95% confidence interval is $10 \pm 0.196 * sqrt(0.1)$, i.e. 9.38 to 10.62.

Example A.5 Theorem A.2 and the Poisson log-likelihood.

According to Theorem A.2, $2\{L(\hat{\mu}) - L(\mu)\}$ follows a chi-square distribution on 1 degree of freedom, from which it follows that $D(\mu) = 2\{L(\hat{\mu}) - L(\mu)\} < 3.84$ with probability 0.95, i.e. the set of values of μ consistent with this inequality also defines a 95% confidence interval for μ. As in Example A.4, set $n = 100$ and $\bar{y} = 10$. Then,

$$D(\mu) = 2\{1302.585 + 100(\mu - 10\log(\mu))\}.$$

It is easy to check by trial-and-error that $D(\mu) < 3.84$ when μ lies between 9.39 and 10.63. This explains the choice $c = 3.84/2 = 1.92$ in Example A.3.

The relevance of Theorems A.1 and A.2 to hypothesis testing is that *the hypothesis $\theta = \theta_0$ is rejected at the $100\alpha\%$ level of significance if θ_0 is not contained in a $100(1 - \alpha)\%$ confidence interval for θ.* In particular, consider the deviance function $D(\theta)$. If θ is a scalar, then the hypothesis $\theta = \theta_0$ is rejected at the 5% level of significance if $D(\theta_0) > 3.84$. If we prefer not to specify the level of significance beforehand, $P(\chi_1^2 > D(\theta_0))$ is called the *p-value* of the test.

When θ is multi-dimensional, the same principles apply, but parameter estimation becomes operationally more complex because a confidence interval becomes a *confidence set* in multi-dimensional space. We can use Theorem A.1 to calculate a confidence set that takes the form of an ellipsoid. In principle, Theorem A.2 can also be used to calculate a confidence set by searching through the parameter space to identify the set of values of θ for which the deviance, $D(\theta)$ is less than the required quantile of the χ_p^2 distribution. In practice,this becomes extremely difficult as the dimensionality, p, increases.

Visualising a confidence set is also difficult in high dimensions. Most statistical software systems limit themselves to reporting estimates and standard errors for individual parameters. Theorem A.1 can then still be used to calculate approximate confidence intervals for each element of θ as the estimated value plus or minus 1.96 times the standard error. However, it is important to recognise that combinations of individual confidence intervals do not generally constitute sensible confidence sets.

The multi-dimensional version of a likelihood ratio test is more straightforward. If θ is a vector of dimension p, then the hypothesis $\theta = \theta_0$ is rejected at the 5% level of significance if $D(\theta_0) > c_p$ where c_p is the 0.95 quantile of the chi-squared distribution on p degrees of freedom, i.e. $P\{\chi_p^2 > c_p\} = 0.05$.

A very useful extension to likelihood-based inference is the method of *profile likelihood*, which operates as follows. Suppose that θ is partitioned as $\theta = (\theta_1, \theta_2)$, with corresponding numbers of elements p_1 and p_2. Suppose also that our primary objective is inference about θ_1. For each possible value of θ_1, let $\hat{\theta}_2(\theta_1)$ be the value of θ_2 which maximises the log-likelihood with θ_1 held fixed. We call $L_P(\theta_1) = L\{\theta_1, \hat{\theta}_2(\theta_1)\}$ the *profile log-likelihood* for θ_1. Then, an extension to Theorem A.2 states that we can treat the profile log-likelihood as if it were a log-likelihood for a model with parameter θ_1 of dimension p_1. Specifically, if we define the deviance function for θ_1 as

$$D(\theta_1) = 2\{L_P(\hat{\theta}_1) - L_P(\theta_1)\}$$

then, asymptotically, $D(\theta_1)$ is distributed as chi-squared on p_1 degrees of freedom. This result provides a method for eliminating the effects of the nuisance parameters θ_2 when making inference about the parameters of interest, θ_1.

The extension of Theorem A.2 is very widely used for comparing *nested models*, i.e. two models one of which is a special case of the other. Suppose we want to test whether a model with $p+q$ parameters gives a better fit to a data-set than a special case of the model in which p of the parameters are set to zero. Let $\hat{\theta}$ be the unrestricted maximum likelihood estimate of θ and $\hat{\theta}_0$ the maximum likelihood constrained by fixing p of the parameters to be zero. Then, if the simpler model is correct,

$$D = 2\{L(\hat{\theta}) - L(\hat{\theta}_0)\} \sim \chi_p^2.$$

A.3.4   Bayesian inference

In Bayesian inference, the likelihood again plays a fundamental role, and θ is again considered as an unknown quantity. However, the crucial difference from classical inference is that θ is considered to be a *random* variable. Hence, the model-specification $[Y|\theta]$ must be converted to a *joint* distribution for Y and θ by specifying a marginal distribution for θ, hence $[Y, \theta] = [Y|\theta][\theta]$. The marginal distribution of θ is also called the *prior* distribution of θ. Its role is to describe the (lack of) knowledge about θ in the absence of the data, Y. The process of inference then consists of asking how conditioning on the realised data, y, changes the prior for θ into its corresponding *posterior* distribution, $[\theta|y]$. The mechanics of this are provided by Bayes' Theorem,

$$[\theta|Y] = [Y|\theta][\theta]/[Y],$$

where $[Y] = \int [Y|\theta][\theta]d\theta$ is the marginal distribution of Y induced by the combination of the specified model, or likelihood function, and the specified prior.

Bayesian inferential statements about θ are expressed as probabilities calculated from the posterior, $[\theta|Y]$. For example, the Bayesian counterpart of a confidence set is a *credible set*, defined as any set $\mathcal{S}$ such that

$$P(\theta \in \mathcal{S}|Y) = \beta,$$

where β is a pre-specified value, conventionally 0.95 by analogy with the conventional use of 95% confidence sets. If a point estimate of θ is required, candidates include the mean or mode of the posterior distribution. Operationally, the crucial requirement for Bayesian inference is the evaluation of the integral which gives the marginal distribution of Y. For many years, this requirement restricted the practical application of Bayesian inference to simple problems. For complex problems and data structures, classical inference involving numerical evaluation and maximisation of the likelihood function was a more practical strategy. However, the situation changed radically with the recognition that Monte Carlo methods of integration, and in particular Markov chain Monte Carlo methods of the kind proposed in Hastings (1970), could be used to generate simulated samples from the posteriors in very complex models. As a result, Bayesian methods are now used in many different areas of application.

A.3.5   Prediction

We now compare classical and Bayesian approaches to prediction. To do so, we need to expand our model specification, $[Y|\theta]$, to include a *target* for prediction, T, which is another random variable. Hence, the model becomes $[T, Y|\theta]$, a specification of the joint distribution of T and Y for a given value of θ. From a classical inferential perspective, we then need to manipulate the model using Bayes' Theorem to obtain the *predictive* distribution for T as the corresponding conditional, $[T|Y, \theta]$. The data give us the realised value of Y, and to complete the predictive inference for T we can either plug-in the maximum likelihood estimate $\hat{\theta}$ or examine how the predictive distribution varies over a range of values of θ contained in its confidence set.

From a Bayesian perspective, the relevant predictive distribution is $[T|Y]$ i.e., the distribution of the target conditional on what has been observed.

Using standard conditional probability arguments, we can express this as

$$[T|Y] = \int [T, \theta|Y]d\theta$$

$$= \int [T|Y, \theta][\theta|Y]d\theta, \qquad (A.11)$$

which shows that the Bayesian predictive distribution is a weighted average of plug-in predictive distributions, with the weights determined by the posterior for θ. A non-Bayesian counterpart to this would be to replace the posterior distribution $[\theta|Y]$ on the right-hand-side of (A.11) by the multivariate Normal distribution of $\hat{\theta}$ given by Theorem A.1. Our earlier comment about transforming a parameter to a scale on which Theorem A.1 gives a more accurate approximation is particularly relevant here.

Note that under either the classical plug-in or the Bayesian approach, the answer to a prediction question is a probability distribution. If we want to summarise this distribution, so as to give a *point prediction*, an obvious candidate summary is the mean i.e., the conditional expectation of T given Y. As discussed in Chapter 2, a theoretical justification for this choice is that it mimimises mean square prediction error. However, we emphasise that in general, the mean is just one of several reasonable summaries of the predictive distribution.

Notice that if θ has a known value, then the manipulations needed for classical prediction of T are exactly the manipulations needed for Bayesian inference treating T as a parameter. It follows that in the Bayesian approach, in which parameters are treated as random variables, the distinction between estimation and prediction is not sharp; from a strictly mathematical point of view, the two are identical. We feel that it is useful to maintain the distinction to emphasise that estimation and prediction address different scientific questions.

A.4   Monte Carlo methods

Monte Carlo methods, also called stochastic simulation, are used routinely in modern statistical methodology, both to conduct statistical experiments and as an inferential tool.

Statistical experiments involve the generation of synthetic samples from known distributions in order to demonstrate that a proposed method has good statistical properties or to conduct sample size calculations. Typically, the experiments use built-in software functionality to generate the required samples, for example in R these include the functions `runif()`, `rnorm()`, `rbinom`, `rpois`, etc. Good general explanations of the underlying theory and methods for stochastic simulation include Morgan (1984) or Ripley (1987)

Statistical experiments are deductive: we know how the "data" were generated and wish to deduce the resulting properties of a particular statistical method. Since the 1970s, Monte Carlo methods have also become increasingly important as inferential tools, where their role is to circumvent problems of analytical tractability.

A.4.1   Direct simulation

We use the term *direct simulation* to mean that we have an algorithm that can generate an independent random sample from the distribution of interest. Suppose, for example, that we wish to use a set of data, Y, to test a null hypothesis $\mathcal{H}_0$ using a test statistic $T = t(Y)$. Let t_0 be the value of T calculated from the observed data. Suppose that the distribution of T under $\mathcal{H}_0$ is intractable, but that we can simulate data Y under $\mathcal{H}_0$. It follows that by repeated simulation of Y we can generate a sample $t_1, t_2, ..., t_b$ of independent realisations of T under $\mathcal{H}_0$, and the proportion of these b values that are greater than t_0 is an unbiased estimate of the p-value of the test. Also, the variance of this estimate is inversely proportional to b, and the Monte Carlo error in the estimated p-value can therefore be made as small as we like by increasing the value of b.

In this book, our main use of direct simulation is in drawing independent random samples from multivariate Normal distributions, which are required for predictive inferences associated with the linear Gaussian model.

A.4.2   Markov chain Monte Carlo

In a complex problem for which a direct simulation algorithm is not available, it may still be possible to draw samples from an approximation to the required distribution using an ingenious device whereby we can construct a sequence of values as a Markov chain whose distribution converges to the required distribution.

A *Markov chain*, $X^t : t = 1, 2, ...$ is a stochastic process with the property that the conditional distribution of X^{t+1} given the complete history, $X^1, ..., X^t$, depends only on X^t. Good introductions to Markov chain Monte Carlo (MCMC) methods include Gilks et al. (1996) and Gamerman & Lopes (2006).

Two examples of Markov chains that illustrate the general idea of MCMC are the *Gibbs sampler* and the *Metropolis algorithm*. In each case, we assume that the goal is to draw a sample of values from an intractable distribution, typically the predictive distribution of a vector-valued target, T, or the posterior distribution of a parameter, θ. In either case, we denote the dimensionality of X by n and write the required distribution as $[X] = [X_1, ..., X_n]$.

The *Gibbs sampler* was named by Geman & Geman (1984) but its use dates back at least to Ripley (1979). It is applicable to problems in which $[X]$ is intractable but it is straightforward to the draw a sample from each of the associated *full conditionals*, $[X_i|\{X_j : j \neq i\}]$. This holds, for example, for Markov random field models Rue & Held (2005). Now, write $X^t = (X_1^t, ..., X_n^t)$. In its simplest form, the Gibbs sampler proceeds as follows.

1. Set $t = 0$ and pick an initial value X^0.

2. For each of $i = 1, ..., n$, generate X_i^{t+1} from the full conditional $[X_i|\{X_j^t : j \neq i\}]$.

3. Increment t to $t + 1$ and repeat step 2.

The marginal distribution of the resulting sequence X^t converges to the required distribution, $[X]$ as t tends to infinity. A useful refinement of the algorithm is to incorporate updated values of X_j^t as soon as they are available. For example, the value of X_2^{t+1} would then be drawn from the full conditional $[X_i|\{X_1^{t=1}, X_3^t, ..., X_n^t\}]$.

The *Metropolis sampler*, due to Metropolis et al. (1998), is applicable to problems in which the required distribution $[X]$ is known up to a multiplicative constant. This typically holds for hierarchically specified models whose distribution is of the form $[Y, S] = [S][Y|S]$ where Y is observed and S is unobserved, for which the likelihood requires integration with respect to S. This includes the class of generalized linear geostatistical models.

The Metropolis algorithm requires the use to specify a *proposal distribution* $[Y|X]$ with the property that $[Y|X] = [X|Y]$. A widely used example of $[Y|X]$ with this property is a multivariate Normal distribution with mean X. The algorithm proceeds as follows.

1. Set $t = 0$ and pick an initial value X^0.

2. Generate Y from the distribution $[Y|X^t]$.

3. Calculate the *acceptance ratio*, $r = \min\{1, [Y]/[X^t]\}$.

4. Generate U from the uniform distribution on $(0, 1)$.

5. If $U < r$, set $X^{t+1} = Y$, otherwise, set $X^{t+1} = X^t$.

6. Increment t to $t + 1$ and repeat steps 2 to 5.

The marginal distribution of the resulting sequence X^t converges to the required distribution, $[X]$ as t tends to infinity. As with the Gibbs sampler, various refinements are possible. One is that the elements of X^t can be updated one component at a time. Another, due to Hastings (1970), does not require the proposal distribution to be symmetric in X and Y.

The Gibbs and Metropolis samplers are but two of many algorithms that have been proposed in a very extensive literature aimed at devising MCMC algorithms that are both computationally and statistically efficient for wide classes of problem. Whatever the algorithm, two critical questions for the user are the following.

1. How many updates to X^t are needed before the algorithm has, for all practical purposes, converged to the required distribution?

2. Once the algorithm has converged, how many samples are needed to make a reliable inferential statement?

The answers to both questions are both algorithm-specific and problem-specific, and can vary by orders of magnitude.

With respect to the first question, Chapter X of Gilks et al. (1996) describes a range of techniques that can be used to monitor the generated sequence X^t. The initial part of the sequence, during which the algorithm is judged not to have converged, is called the *burn-in* period of the chain and is discarded. The remainder of the sequence is then assumed to be a sample from the required distribution. One superficially simple test is to compare the empirical distributions of X^t from the first and second halves of the post-burn-in sequence – "superficially" simple because of the stochastic dependence amongst the X^t, which we now discuss.

The second question is difficult to answer in full generality because successive values of the post-burn-in sequence X^t are stochastically dependent, often very strongly so; as an extreme example, samples generated by a Metropolis algorithm include sub-samples of identical values whenever step 5 of the algorithm sets $X^{t+1} = X^t$. One consequence of this is that the observed proportion out of n sampled values that meet a condition of interest, such as exceedance of a specified threshold, is an unbisased estimate of the corresponding probability, p, but the variance, v, of this estimate can be very much bigger than the formula applicable to an independent random sample, $v = p(1-p)/n$. It is common practice to thin the post-burn-in output of an MCMC algorithm, i.e. retain only every kth sampled value. This weakens the dependence between successive retained values, leading to better precision for a given size of *retained* sample. Estimates based on a thinned sample cannot be as precise as those based on their unthinned counterparts, but when storage space is at a premium some sacrifice of precision may be worthwhile.

Most importantly of all, whilst MCMC is a powerful tool that has had an enormous impact on statistical practice, in any specific application it is the user's responsibility to satisfy themself and their audience that the associated inferences are substantially correct.

A.4.3   Monte Carlo maximum likelihood

We consider the class of probabilistic models defined as follows.

1. $S = (S_1, \ldots, S_n)$ follows a multivariate Gaussian distribution with mean μ and covariance matrix Σ.

2. Conditionally on S, $Y = (Y_1, \ldots, Y_n)$ are mutually independent random variables whose joint distribution is of the form

$$[Y|S] = \prod_{i=1}^{n} [Y_i|S_i].$$

We denote by θ the complete set of parameters that together identify the joint distribution of S and Y. Hence, θ includes parameters that determine μ and Σ together with any additional parameters of the conditional distributions $[Y_i|S_i]$.

The likelihood for θ given a set of observed values $y = (y_1, \ldots, y_n)$ follows by integrating out the unobserved random variables S, to give

$$L(\theta) = \int [S;\theta][y|S]\, dS$$

$$= \int [S;\theta] \prod_{i=1}^{n} [y_i|S_i;\theta]\, dS \qquad (A.12)$$

where the extended notation $[S;\theta]$ indicates that the multivariate Gaussian distribution of S involves the set of parameters θ, and similarly for $[y_i|S_i;\theta]$. Typically, the integral on the right-hand side of (A.12) cannot be expressed in closed form. Monte Carlo methods then provide a way of approximating the likelihood function, as follows.

Let θ_0 be an initial guess for the value of the maximum likelihood estimate, $\hat{\theta}$. Now, re-express (A.12) as

$$L(\theta) = \int [S;\theta][y|S] \times \frac{[S,y;\theta_0]}{[S,y;\theta_0]}\, dS$$

$$= \int \frac{[S;\theta][y|S]}{[S;\theta_0][y|S]} \times [S,y;\theta_0]\, dS$$

$$= \int \frac{[S;\theta]}{[S;\theta_0]} \times [S,y;\theta_0]\, dS$$

$$\propto \int \frac{[S;\theta]}{[S;\theta_0]} \times [S|y;\theta_0]\, dS$$

By simulating B samples, say $S_{(b)} : b = 1, ..., B$, from $[S|y;\theta_0]$ we then approximate the integral in (A.13) with

$$L_B(\theta) = \frac{1}{B} \sum_{b=1}^{B} \frac{[S_{(b)};\theta]}{[S_{(b)};\theta_0]}.$$

Finally, we maximize $L_B(\theta)$ with respect to θ to obtain an estimate $\hat{\theta}_B$. This is a *Monte Carlo maximum likelihood* estimate for θ. To simulate from $[S|y;\theta_0]$, Markov Chain Monte Carlo algorithms (see Section A.4.2) can be used whenever direct simulation from $[S|y;\theta_0]$ is not feasible.

The performance of $\hat{\theta}_B$ depends on the value of B, which should be large enough to make the Monte Carlo error negligible relative to the inherent statistical imprecision of $\hat{\theta}_B$. Also, and critically, the quality of the approximation to the maximum likelihood estimate $\hat{\theta}$ depends on how close θ_0 is to $\hat{\theta}$. For this reason, it is both legitimate and desirable to consider $\hat{\theta}_B$ as a provisional estimate, set $\theta_0 = \hat{\theta}_B$ and repeat the process with a larger value of B to check that this makes no material difference to the result.

B

Spatial data handling

CONTENTS

Here, we provide a brief overview of how to handle spatial data in R so as to produce high-quality maps. We use Liberia as an example. The required packages are: `sf`, `raster`, `leaflet` and `tmap`.

To install a package, execute the R command

```
install.packages(''TYPE NAME OF THE PACKAGE'')
```

In the remainder of this Appendix, we consider the two main types of spatial data, namely *vectors* and *rasters*, and show how to create static and interactive maps in R. The spatial data for Liberia are freely available for download at the following link

http://www.diva-gis.org/Data

The reader may wish to replicate the results for other countries whose data are available from the above link.

B.1 Handling vector data in R

A geographical *vector* data-set, also commonly called a *shapefile*, consists of a set of geographical points within a coordinate reference system (CRS) representing stand-alone objects (e.g. the location of a specific place such as a hospital, school or household) or more complex interconnected features taking the form of lines and polygons (e.g. roads, rivers and lakes). This concept is not exclusive to R and should not be confused with the objects of the R `vector` class.

A geographical *vector* consists of three mandatory files having the following extensions: `.shp`, containing the geometry of the object to represent; `.shx`, a compiled version of the geometry in the `.shp` file to allow for faster searches; `.dbf`, giving the attributes of each shape (e.g. the names of districts which partition a region). A fourth file, with the `.prj` extension, gives information on the CRS and the projection used to represent the geometry.

We now consider *vector* data representing the second level administrative subdivision of Liberia. The data are contained in the files LBR_adm2.shp, LBR_adm2.shx, LBR_adm2.dbf and LBR_adm2.prj. The function st_read reads this *vector* data by inputting any of the three mandatory files using the command

```
> Liberia.adm2 <- st_read("LBR_adm2.shp")
```

To access the CRS of this file, we type

```
> st_crs(Liberia.adm2)
Coordinate Reference System:
  EPSG: 4326
  proj4string: "+proj=longlat+datum=WGS84+no_defs"
```

The EPSG code 4326 identifies a WGS84 longitude/latitude CRS. The WGS84 is the most commonly used CRS, and is the default option for almost all *vector* data (and, incidentally, for *raster* data).

For geostatistical analysis, it is often convenient to project the available geographical information onto a two-dimensional Cartesian CRS. This is a mandatory step if the distance between two spatial objects needs to be computed. To achieve this, a commonly used CRS is the Universal Transverse Mercator (UTM), which divides the earth into 60 longitudinal wedges and 20 latitudinal segments. To re-project our *vector*, we first need to find the corresponding UTM zone and its corresponding EPSG code. For Liberia, this is 32629, indicating the 29-th UTM zone north of the equator. The transformation of the geometry into the new CRS is obtained as follows

```
Liberia.adm2.utm <- st_transform(Liberia.adm2,32629)
```

We then check the CRS with

```
> st_crs(Liberia.adm2.utm)
Coordinate Reference System:
  EPSG: 32629
  proj4string: "+proj=utm+zone=29+datum=WGS84
+units=m +no_defs"
```

To obtain the geometry of the borders of Liberia, excluding those of the administrative subdivision, we type

```
> Liberia.union <- st_union(Liberia.adm2)
```

When carrying out spatial prediction, we often need to create a regular grid covering the whole of the study area. The code below gives an example on how to achieve this by creating a regular grid within the boundaries of Liberia.

```
> Liberia.grid.sq <- st_make_grid(Liberia.adm2.utm,
+                                   cellsize = 5000,
```

```
+                                    what="centers")
>
> Liberia.inout <- st_intersects(Liberia.grid.sq,
+                                 Liberia.union,
+                                 sparse = FALSE)
> Liberia.grid <- Liberia.grid.sq[Liberia.inout]
```

Using the function `st_make_grid`, we created a 5 by 5 km regular grid (specified by `cellsize=5000`) within a box containing Liberia. The function `st_intersects` then returns a logical vector indicating whether each location in `Liberia.grid.sq` falls within the polygon `Liberia.union` or not. Finally, we use `Liberia.inout` to take the subset of locations in `Liberia.grid.sq` that fall within Liberia.

In order to create a variable whose entries are associated with each polygon of the subdivision of Liberia, the following code gives an example for the area in square meters of each district

```
> Liberia.adm2.utm$Area <- st_area(Liberia.adm2.utm)
```

To display a map of Liberia, we use the packages `tmap` and `leaflet`. A handy feature of the `tmap` package is that it allows the creation of ''tmap'' objects containing maps that can be later re-used to add more features.

The code below creates an object called `Map.adm2.border`

```
> Map.adm2.border <-
+   tm_shape(Liberia.adm2) +
+   tm_borders()
```

We now use this object to create three maps.

```
> map1 <- Map.adm2.border
>
> map2 <- Map.adm2.border+
+   tm_shape(Liberia.union)+tm_borders(lwd=2,col="red")+
+   tm_shape(Liberia.grid)+tm_dots(col="blue")
>
> map3 <- Map.adm2.border+
+   tm_fill("Area")+
+   tm_compass(type="8star",
+              position = c("right","top"))+
+   tm_scale_bar(breaks = c(0,100,200),size=1,
+                position=c("center","bottom"))
>
> tmap_arrange(map1,map2,map3,ncol=3)
```

The resulting set of maps is shown in Figure B.1. The `tmap_arrange` function above provides an easy way to arrange multiple maps in a single figure. Also, note that in the third map, `map3`, we have added a scale bar with `tm_scale_bar`, and a compass with `tm_compass`.

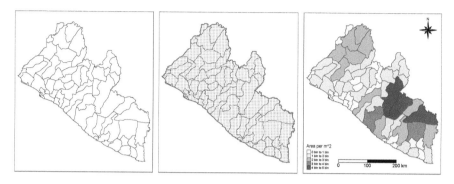

FIGURE B.1
The left panel shows a map of the second level subdivision of Liberia. The central and right panels add a 5 by 5 regular grid and the area of each district, respectively.

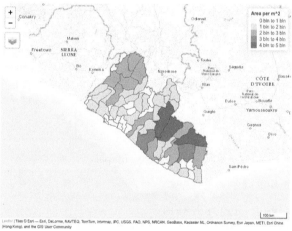

FIGURE B.2
Interactive map of the area of each district in Liberia obtained using the **tmap** package.

To add more context to a map, we can use the interactive functionalities provided by the **tmap** package.

```
> tmap_mode("view")
tmap mode set to interactive viewing
> map3
```

The first line of the code above turns on the interactive features; to switch back to static maps, enter **tmap_map("plot")**. The interactive display allows us to zoom in and out to any desired area of the map (see Figure B.2). This can be particularly useful in order to better visualize the values of a variable for small districts.

B.2 Handling raster data in R

In this section we consider the simplest form of a *raster* data-set, which consists of a matrix whose entries, referred to as pixels, contain the values of a real-valued spatially referenced variable, e.g. temperature, elevation or population density.

Using the **raster** package, a raster file can be loaded in R as follows

```
> elev <- raster("LBR_alt.gri")
```

The above object, `elev`, is a raster of the elevation in meters for Liberia. We then transform this into the correct UTM projection

```
> elev <- projectRaster(elev,
+          crs="+init=epsg:32629")
```

We then create a static map of the raster in **tmap**

```
> map.elev <- tm_shape(elev)+tm_raster(title="Elevation")+
+    tm_shape(Liberia.union)+tm_borders(col="black")
> map.elev
```

To crop the elevation map so that this it is shown only within the boundaries of Liberia, we use the **mask** function

```
> Liberia.elev <- mask(elev,
+                 as(Liberia.union,"Spatial"))
```

We then visualise the cropped map as

```
> map.Liberia.elev <- tm_shape(Liberia.elev)+
+ tm_raster(title="Elevation")+
+                 tm_shape(Liberia.union)+tm_borders(col="black")
> map.Liberia.elev
```

The result of this operation is shown in Figure B.3.

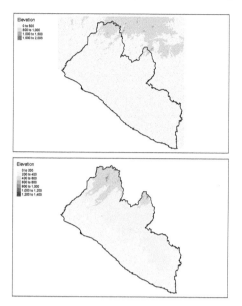

FIGURE B.3
Maps of the elevation, in meters, in Liberia.

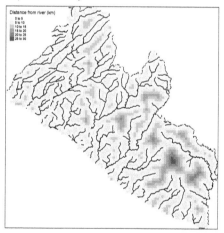

FIGURE B.4
Map of the distance from the closest river on a 5 by 5 km regular grid. The black lines represent the digitized rivers contained in the **water** object.

To extract the value of elevation at a set of predefined locations, we use the function **extract**

```
> sample.coords <- st_sample(Liberia.union,size=5)
```

```
> extract(elev,as(sample.coords,"Spatial"))
[1]   79.31655  217.86434  352.00581  393.19660    52.00399
```

In the code above, the object `sample.coords` contains five locations that are randomly sampled within the boundaries of Liberia and for which we then print their corresponding elevations.

For many vector-borne diseases, proximity to waterways is an important risk factor. Using the 5 by 5 km grid over Liberia (see `Liberia.grid` above), we create a raster file containing the distance of each pixel from the closest waterway, measured in kilometres, as follows

```
> water <- st_read("LBR_water_lines_dcw.shp")
>
> water <- st_transform(water,32629)
>
> dist <- apply(st_distance(Liberia.grid,water),1,min)/1000
> dist.raster <- rasterFromXYZ(cbind(
+                   st_coordinates(Liberia.grid),
+                   dist),
+                   crs="+init=epsg:32629")
> tm_shape(dist.raster)+
+ tm_raster(title="Distance␣from␣river␣(km)")+
+ tm_shape(water)+tm_lines()
```

In the above code, the file `water` is a *vector* data-set of waterways in Liberia. The function `st_distance` computes the minimum distance between each point on the grid and the closest river. The `rasterFromXYZ` function creates the raster object `dist.water`, which is displayed in Figure B.4.

References

Amek, N., Bayoh, N., Hamel, M., Lindblade, K. A., Gimnig, J., Laserson, K. F., ... Vounatsou, P. (2011). Spatio-temporal modeling of sparse geo-statistical malaria sporozoite rate data using a zero inflated binomial model. *Spatial and Spatio-temporal Epidemiology, 2*, 283-290.

Amoah, B., Giorgi, E., & Diggle, P. J. (2018). *A geostatis-tical framework for combining spatially referenced disease prevalence data from multiple diagnostics.* (Submitted. Pre-print available at https://arxiv.org/abs/1808.03141)

ANSD. (2015). *Séńgal : Enquête démographique et de santćontinue (eds-continue 2014).* Rockville, Maryland, USA : Agence Nationale de la Statistique et de la Démographie and ICF International.

Armstrong, M. P., Rushton, G., & Zimmerman, D. L. (1999). Geographically masking health data to preserve confidentiality. *Statistics in Medicine, 18*, 497–525.

Azzalini, A. (1996). *Statistical Inference: Based on the Likelihood.* Boca Raton: Chapman and Hall/CRC Press.

Baddeley, A., Rubak, E., & Turner, R. (2016). *Spatial Point Patterns: Methodology and Applications with R.* Boca Raton: Chapman and Hall/CRC Press.

Banerjee, S., Gelfand, A., Finley, A., & Sang, H. (2008). Gaussian predictive process models for large spatial data sets. *Journal of the Royal Statistical Society, Series B, 70*, 825–848.

Bennett, A., Kazembe, L., Mathanga, D., Kinyoki, D., Ali, D., Snow, R., & Noor, A. M. (2013). Mapping malaria transmission intensity in Malawi, 2000-2010. *American Journal of Tropical Medicine and Hygiene, 89*, 840-849.

Berman, M., & Diggle, P. (1989). Estimating weighted integrals of the second-order intensity of a spatial point process. *Journal of the Royal Statistical Society, Series B, 51*, 81–92.

Besag, J. (1974). Spatial interaction and the statistical analysis of lattice systems (with discussion). *Journal of the Royal Statistical Society, Series B, 36*, 192-236.

Besag, J. (1981). On a system of two-dimensional recurrence equations. *Journal of the Royal Statistical Society, Series B*, *43*, 302-309.

Besag, J., York, J., & Mollié, A. (1991). Bayesian image restoration, with two applications in spatial statistics (with discussion). *Annals of the Institute of Statistical Mathematics*, *43*, 1–59.

Bevilacqua, M., & Gaetan, C. (2015). Comparing composite likelihood methods based on pairs for spatial gaussian random fields. *Statistics and Computing*, *25*, 877–892.

Bijleveld, A., van Gils, J., van der Meer, J., Dekinga, A., Kraan, C., van der Veer, H., & Piersma, T. (2012). Designing a benthic monitoring programme with multiple conflicting objectives. *Methods in Ecology and Evolution*, *3*, 526–536.

Box, G., & Cox, D. (1964). An analysis of transformations. *Journal of the Royal Statistical Society, Series B*, *26*, 211–252.

Breslow, N., & Clayton, D. (1993). Approximate inference in generalized linear mixed models. *Journal of the American Statistical Association*, *88*, 9–25.

Chilès, J.-P., & Delfiner, P. (2016). *Geostatistics* (second ed.). Hoboken: Wiley.

Chipeta, M., Terlouw, D., Phiri, K., & Diggle, P. (2016). Adaptive geostatistical design and analysis for prevalence surveys. *Spatial Statistics*, *15*, 70–84.

Chipeta, M., Terlouw, D., Phiri, K., & Diggle, P. (2017). Inhibitory geostatistical designs for spatial prediction taking account of uncertain covariance structure. *Environmetrics*, *28*, e2425. doi: 10.1002/env.2425

Christensen, O. F., Roberts, G. O., & Sköld, M. (2006). Robust Markov chain Monte Carlo methods for spatial generalized linear mixed models. *Journal of Computational and Graphical Statistics*, *15*, 1-17.

Clayton, D., & Hills, M. (1993). *Statistical Models in Epidemiology*. Oxford: Oxford University Press.

Clements, A., Lwambo, N., Blair, L., Nyandindi, U., Kaatano, G., Kinung'hi, S., ... Brooker, S. (2006). Bayesian spatial analysis and disease mapping: tools to enhance planning and implementation of a schistosomiasis control programme in Tanzania. *Tropical Medicine and International Health*, *11*, 490-503.

Coffeng, L., Stolk, W., Zoure, H., Veerman, J., Agblewonu, K., Murdoch, M., ... Amazigo, U. (2013). African Programme for Onchocerciasis Control 1995-2015: model-estimated health impact and cost. *PLoS Neglected Tropical Diseases*, *7*. doi: e2032. doi:10.1371/journal.pntd.0002032

Corran, P., Coleman, P., Riley, E., & Drakeley, C. (2007). Serology: a robust indicator of malaria transmission intensity? *Trends in Parasitology*, *23*, 575-582.

Cox, D. (1977). The role of significance tests. *Scandinavian Journal of Statistics*, *4*, 49–71.

Cox, D., & Hinkley, D. (1974). *Theoretical Statistics*. London: Chapman and Hall.

Cressie, N. (1991). *Statistics for Spatial Data*. New York: Wiley.

Daniels, M., & Hogan, J. (2008). *Missing Data in Longitudinal Studies: Strategies for Bayesian Modeling and Sensitivity Analysis*. Boca Raton: Chapman and Hall/CRC.

De Boor, C. (2001). *A Practical Guide to Splines*. New York: Springer.

Diggle, P. (1983). *Statistical Analysis of Spatial Point Patterns*. London: Academic Press.

Diggle, P. (2013). *Statistical Analysis of Spatial and Spatio-Temporal Point Patterns* (third ed.). Boca Raton: Chapman and Hall/CRC Press.

Diggle, P., & Giorgi, E. (2016). Model-based geostatistics for prevalence mapping in low-resource settings (with discussion). *Journal of the American Statistical Association*, *111*, 1096–1120.

Diggle, P., & Kenward, M. (1994). Informative dropout in longitudinal data analysis (with discussion). *Applied Statistics*, *43*, 49–93.

Diggle, P., & Lophaven, S. (2006). Bayesian geostatistical design. *Scandinavian Journal of Statistics*, *33*, 53–64.

Diggle, P., Menezes, R., & Su, T.-L. (2010). Geostatistical analysis under preferential sampling (with discussion). *Applied Statistics*, *59*, 191–232.

Diggle, P., Moraga, P., Rowlingson, B., & Taylor, B. (2013). Spatial and spatio-temporal log-gaussian cox processes: extending the geostatistical paradigm. *Statistical Scioence*, *28*, 542–563.

Diggle, P., Moyeed, R., & Tawn, J. (1998). Model-based geostatistics (with discussion). *Applied Statistics*, *47*, 299–350.

Diggle, P., & Ribeiro, P. J. (2007). *Model-based Geostatistics*. Springer Science+Business Media, New York.

Dobson, A. J., & Barnett, A. (2008). *An Introduction to Generalized Linear Models* (third ed.). Chapman and Hall/CRC.

Fanshawe, T. R., & Diggle, P. J. (2011). Spatial prediction in the presence of positional error. *Environmetrics*, *22*, 109–122.

Firth, D. (1988). Multiplicative errors: Log-normal or gamma? *Journal of the Royal Statistical Society. Series B (Methodological)*, *50*, 266-268. Retrieved from http://www.jstor.org/stable/2345764

Fong, Y., Rue, H., & Wakefield, J. (2010). Bayesian inference for generalized linear mixed models. *Biostatistics*, *11*, 397. Retrieved from + http://dx.doi.org/10.1093/biostatistics/kxp053 doi: 10.1093/biostatistics/kxp053

Fronterrè, C., Giorgi, E., & Diggle, P. (2018). Geostatistical inference in the presence of geomasking: A composite-likelihood approach. *Spatial Statistics*, *28*, 319–330.

Gamerman, D., & Lopes, H. (2006). *Markov Chain Monte Carlo: Stochastic Simulation for Bayesian Inference* (second ed.). Boca Raton: Chapman and Hall/CRC Press.

Gelfand, A., Diggle, P., Fuentes, M., & Guttorp, P. (2010). *A Handbook of Spatial Statistics*. Boca Raton: Chapman and Hall/CRC Press.

Geman, S., & Geman, D. (1984). Stochastic relaxation, gibbs distributions and the Bayesian restoration of images. *IEEE Transactions in Pattern Analysis and Machne Intelligence*, *6*, 721–741.

Gething, P. W., Elyazar, I. R. F., Moyes, C. L., Smith, D. L., Battle, K. E., Guerra, C. A., ... Hay, S. I. (2012). A long neglected world malaria map: *Plasmodium vivax* endemicity in 2010. *PLoS Neglected Tropical Diseases*, *6*, e1814. doi: 10.1371/journal.pntd.0001814

Giardina, F., Gosoniu, L., Konate, L., Diouf, M. B., Perry, R., Gaye, O., ... Vounatsou, P. (2012). Estimating the burden of malaria in Senegal: Bayesian zero-inflated binomial geostatistical modeling of the MIS 2008 data. *PLoS ONE*, *7*, e32625. doi: 10.1371/journal.pone.0032625

Gilks, W., Richardson, S., & Spiegelhalter, D. (1996). *Markov Chain Monte Carlo in Practice*. Boca Raton: Chapman and Hall/CRC Press.

Giorgi, E., Diggle, P., Snow, R., & Noor, A. (2018). Geostatistical methods for disease mapping and visualization using data from spatio-temporally referenced prevalence surveys. *International Statistical Review*, *86*, 571597.

Giorgi, E., & Diggle, P. J. (2015). On the inverse geostatistical problem of inference on missing locations. *Spatial Statistics*, *11*, 35 - 44.

Giorgi, E., & Diggle, P. J. (2017). PrevMap: An R package for prevalence mapping. *Journal of Statistical Software*, *78*, 1–29. doi: 10.18637/jss.v078.i08

Giorgi, E., Schlüter, D. K., & Diggle, P. J. (2017). Bivariate geostatistical modelling of the relationship between Loa loa prevalence and intensity of infection. *Environmetrics*, *29*, e2447. doi: 10.1002/env.2447

Giorgi, E., Sesay, S. S. S., Terlouw, D. J., & Diggle, P. J. (2015). Combining data from multiple spatially referenced prevalence surveys using generalized linear geostatistical models. *Journal of the Royal Statistical Society, Series A*, *178*, 445-464.

Gneiting, T. (2002). Nonseparable, stationary covariance functions for space-time data. *Journal of the American Statistical Association*, *97*, 590-600.

Hansell, A. L., Beale, L. A., Ghosh, R. E., Fortunato, L., Fecht, D., Järup, L., & Elliott, P. (2014). *The Environment and Health Atlas for England and Wales*. Oxford University Press.

Hastings, W. (1970). Monte Carlo sampling methods using Markov chains and their applications. *Biometrika*, *57*, 97–109.

Hay, S. I., Guerra, C. A., Gething, P. W., Patil, A. P., Tatem, A. J., Noor, A. M., ... Snow, R. W. (2009). A world malaria map: *Plasmodium falciparum* endemicity in 2007. *PLoS Medicine*, *6*, e1000048. doi: 10.1371/journal.pmed.1000048

Hedt, B. L., & Pagano, M. (2011). Health indicators: Eliminating bias from convenience sampling estimator. *Statistics in Medicine*, *30*, 560-568.

Higdon, D. (1998). A process-convolution approach to modeling temperatures in the North Atlantic Ocean. *Environmental and Ecological Statistics*, *5*, 173-190.

Hogan, J., & Laird, N. (1997). Model-based approaches to analyzing incomplete longitudinal and failure-time data. *Statistics in Medicine*, *16*.

Hurn, M. A., Husby, O. K., & Rue, H. (2003). A tutorial on image analysis. In J. Møller (Ed.), *Spatial Statistics and Computational Methods* (pp. 87–141). New York, NY: Springer New York.

Ilian, J., Penttinen, A., Stoyan, H., & Stoyan, D. (2008). *Statistical Analysis and Modelling of Spatial Point Patterns*. Chichester: Wiley.

Irvine, M. A., Njenga, S. M., Gunawardena, S., Njeri Wamae, C., Cano, J., Brooker, S. J., & Deirdre Hollingsworth, T. (2016). Understanding the relationship between prevalence of microfilariae and antigenaemia using a model of lymphatic filariasis infection. *Transactions of The Royal Society of Tropical Medicine and Hygiene*, *110*, 118–124.

Jardim, E., & Ribeiro, P. (2007). Geostatistical assessment of sampling designs for Portuguese bottom trawl surveys. *Fisheries Resarch*, *85*.

Joe, H. (2008). Accuracy of Laplace approximation for discrete response mixed models. *Computational Statistics & Data Analysis*, *52*, 5066-5074.

Kleinschmidt, I., Pettifor, A., Morris, N., MacPhail, C., & Rees, H. (2007). Geographic distribution of human immunodeficiency virus in South Africa. *American journal of tropical medicine and hygiene*, *77*, 1163-1169.

Kleinschmidt, I., Sharp, B. L., Clarke, G. P. Y., Curtis, B., & Fraser, C. (2001). Use of generalized linear mixed models in the spatial analysis of small-area malaria incidence rates in Kwazulu Natal, South Africa. *American Journal of Epidemiology*, *153*, 1213-1221.

Krige, D. (1951). A statistical approach to some basic mine valuation problems on the witwatersrand. *Journal of the Chemical, Metallurgical and Mining Society of South Africa*, *52*, 119–139.

Langsaeter, A. (1926). Om beregnin av middlefeilen ved regelmessige linjetakseringer. *Meddelanden fra det Norske Skogsforsóksvesen*, *2*, 5–47.

Lark, R. (2002). Optimized spatial sampling of soil for estimation of the variogram by maximum likelihood. *Geoderma*, *105*, 49–80.

Lee, P. (2012). *Bayesian Inference: an Introduction* (fourth ed.). Chichester: Wiley.

Lin, H., Scharfstein, D., & Rosenheck, R. (2004). Analysis of longitudinal data with irregular, outcome-dependent follow-up. *Journal of the Royal Statistical Society, Series B*, *66*, 791–813.

Lindgren, F., & Rue, H. (2015). Bayesian spatial modelling with R-INLA. *Journal of Statistical Software, Articles*, *63*, 1–25. doi: 10.18637/jss.v063.i19

Lindgren, F., Rue, H., & Lindstrom, J. (2011). An explicit link between Gaussian fields and Gaussian Markov random fields: the stochastic partial differential equation approach (with discussion). *Journal of the Royal Statistical Society, Series B*, *73*.

Little, R. (1995). Modelling the drop-out mechanism in repeated-measures studies. *Journal of the American Statistical Association*, *90*, 1112–1121.

López-Abente, G., Ramis, R., Pollán, M., Aragonés, N., Pérez-Gómez, B., Gómez-Barroso, D., ... Garćia-Mendizábal, M. J. (2006). *Atlas municipal de mortalidad por cancer en españa, 1989-1998*. Instituto de Salud Carlos III, Madrid.

Lucy, D., Aykroyd, R. G., & Pollard, A. M. (2002). Nonparametric calibration for age estimation. *Journal of the Royal Statistical Society, Series C*, *51*, 183-196. doi: 10.1111/1467-9876.00262

Matérn, B. (1960). *Spatial variation*. Meddelanden fran Statens Skogsforsknings institut, Stockholm. Band 49, number 5.

Matérn, B. (1986). *Spatial Variation* (second ed.). Berlin: Springer-Verlag.

Matheron, G. (1963). Principles of geostatistics. *Economic Geology, 58*, 1246–1266.

McBratney, A., & Webster, R. (1981). The design of optimal sampling schemes for local estimation and mapping of regionalized variables - II. Program and examples. *Computers and Geosciences, 7*, 335–365.

McBratney, A., Webster, R., & Burgess, T. (1981). The design of optimal sampling schemes for local estimation and mapping of of regionalized variables - i: Theory and method. *Computers and Geosciences, 7*, 331–334.

Metropolis, N., Rosenbluth, A., Rosenbluth, M., Teller, A., & Teller, E. (1998). Equations of state calculations by fast computing machine. *Journal of Chemical Physics, 21*, 1087–1091.

Mogeni, P., Williams, T. N., Omedo, I., Kimani, D., Ngoi, J. M., Mwacharo, J., . . . others (2017). Detecting malaria hotspots: A comparison of rapid diagnostic test, microscopy, and polymerase chain reaction. *The Journal of infectious diseases, 216*, 1091–1098.

Møller, J., Syversveen, A., & Waagepetersen, R. (1998). Log Gaussian Cox processes. *Scandinavian Journal of Statistics, 25*, 451–482.

Morgan, B. (1984). *Elements of simulation*. London: Chapman and Hall.

Müller, W. (2007). *Collecting Spatial Data: Optimum Design of Experiments for Random Fields* (third ed.). Berlin: Springer-Verlag.

Müller, W., & Zimmerman, D. (1999). Optimal designs for variogram estimation. *Enviromentrics, 10*, 23–37.

Neal, R. M. (2011). MCMC using Hamiltonian dynamics. In S. Brooks, A. Gelman, G. Jones, & X.-L. Meng (Eds.), *Handbook of Markov chain Monte Carlo* (p. 113-162). Chapman & Hall, CRC Press.

Nelder, J., & Wedderburn, R. (1972). Generalized linear models. *Journal of the Royal Statistical Society, Series A, 135*, 370–384.

Noor, A. M., Kinyoki, D. K., Mundia, C. W., Kabaria, C. W., Mutua, J. W., Alegana, V. A., . . . Snow, R. W. (2014). The changing risk of *Plasmodium falciparum* malaria infection in Africa: 2000-10: a spatial and temporal analysis of transmission intensity. *The Lancet, 383*, 1739 - 1747. doi: http://dx.doi.org/10.1016/S0140-6736(13)62566-0

Nowak, W. (2010). Measures of parameter uncertainty in geostatistical estimation and geostatistical optimal design. *Mathematical Geosciences, 42,* 199–221.

O'Hagan, A. (1994). *Bayesian Inference.* London: Edward Arnold.

Oluwole, A. S., Ekpo, U. F., Karagiannis-Voules, D.-A., Abe, E. M., Olamiju, F. O., Isiyaku, S., ... Vounatsou, P. (2015). Bayesian geostatistical model-based estimates of soil-transmitted helminth infection in Nigeria, including annual deworming requirements. *PLoS Negl Trop Dis, 9,* e0003740. doi: 10.1371/journal.pntd.0003740

Pati, D., Reich, B., & Dunson, D. (2011). Bayesian geostatistical modelling with informative sampling locations. *Biometrika, 98,* 35–48.

Pawitan, Y. (2001). *In All Likelihood: Statistical Modelling and Inference Using Likelihood.* Oxford: Oxford University Press.

Pettitt, A., & McBratney, A. (1993). Sampling designs for estimating spatial variance components. *Applied Statistics, 42,* 185–209.

Pullan, R. L., Gething, P. W., Smith, J. L., Mwandawiro, C. S., Sturrock, H. J. W., Gitonga, C. W., ... Brooker, S. (2011). Spatial modelling of soil-transmitted helminth infections in Kenya: A disease control planning tool. *PLoS Neglected Tropical Diseases, 5,* e958. doi: 10.1371/journal.pntd.0000958

Raso, G., Matthys, B., N'goran, E. K., Tanner, b., Vounatsou, P., & Utzinger, J. (2005). Spatial risk prediction and mapping of schistosoma mansoni infections among schoolchildren living in western Côte d'Ivoire. *Parasitology, 131,* 97-108. doi: 10.1017/S0031182005007432

Rice, S. O. (1944). Mathematical analysis of random noise. *Bell System Technical Journal, 23,* 282–332.

Ripley, B. (1977). Modelling spatial patterns (with discussion). *Journal of the Royal Statistical Society, Series B, 39,* 172–212.

Ripley, B. (1979). Simulating spatial patterns. *Applied Statistics, 28,* 109–112.

Ripley, B. (1981). *Spatial Statistics.* New York: Wiley.

Ripley, B. (1987). *Stochastic Simulation.* New York: Wiley.

Ritter, K. (1996). Asymptotic optimality of regular sequence designs. *Annals of Statistics, 24,* 2081–2096.

RStudio, Inc. (2013). Easy web applications in r. [Computer software manual]. (http://www.rstudio.com/shiny/)

Rue, H., & Held, L. (2005). *Gaussian Markov Random Fields: Theory and Applications.* London: CRC Press.

Rue, H., Martino, S., & Chopin, N. (2009). Approximate bayesian inference for latent Gaussian models by using integrated nested Laplace approximations. *Journal of the Royal Statistical Society, Series B, 71,* 319–392.

Russo, D. (1984). Design of an optimal sampling network for estimating the variogram. *Soil Science Society of America Journal, 48,* 708–716.

Schlather, M., Ribeiro, P., & Diggle, P. (2004). Detecting dependence between marks and locations of marked point processes. *Journal of the Royal Statistical Society, Series B, 66,* 79–93.

Schlüter, D. K., Ndeffo-Mbah, M. L., Takougang, I., Ukety, T., Wandji, S., Galvani, A. P., & Diggle, P. J. (2016). Using community-level prevalence of *loa loa* infection to predict the proportion of highly-infected individuals: Statistical modelling to support lymphatic filariasis and onchocerciasis elimination programs. *PLoS Neglected Tropical Diseases., 10.* doi: 10.1371/journal.pntd.0005157

Self, S., & Liang, K.-Y. (1987). Asymptotic properties of maximum likelihood estimators and likelihood ratio tests under nonstandard conditions. *Journal of the American Statistical Association, 82,* 605–610.

Shepard, D. (1968). A two-dimensional interpolation function for irregularly-spaced data. *Proceedings of the 1968 ACM national conference,* 517?524. doi: 10.1145/800186.810616

Snow, R., Amratia, P., Mundia, C., Alegana, V., Kirui, V., Kabaria, C., & Noor, A. (2015). *Assembling a geo-coded repository of malaria infection prevalence survey data in Africa 1900-2014.* (Tech. Rep.). (INFORM Working Paper, developed with support from the Department of International Development and Wellcome Trust, UK, June 2015.)

Snow, R. W., Kibuchi, E., Karuri, S. W., Sang, G., Gitonga, C. W., Mwandawiro, C., ... Noor, A. M. (2015). Changing malaria prevalence on the kenyan coast since 1974: Climate, drugs and vector control. *PLoS ONE, 10,* 1-14. doi: 10.1371/journal.pone.0128792

Soares Magalhaes, R. J., & Clements, A. C. A. (2011). Mapping the risk of anaemia in preschool-age children: The contribution of malnutrition, malaria, and helminth infections in West Africa. *PLoS Medicine, 8,* e1000438. doi: 10.1371/journal.pmed.1000438

Stein, M. (1999). *Interpolation of Spatial Data: Some Theory for Kriging.* New York: Springer-Verlag.

Stein, M. L. (2005). Space time covariance functions. *Journal of the American Statistical Association, 100,* 310-321.

Stein, M. L., Chi, Z., & Welty, L. J. (2004). Approximating likelihoods for large spatial data sets. *Journal of the Royal Statistical Society, Series B*, *66*, 275–296.

Stevenson, J. C., Stresman, G. H., Gitonga, C. W., Gillig, J., Owaga, C., Marube, E., ... Cox, J. (2013). Reliability of school surveys in estimating geographic variation in malaria transmission in the Western Kenyan highlands. *PLoS ONE*, *8*, e77641. doi: 10.1371/journal.pone.0077641

Stigler, S. M. (1980). Stigler's law of eponomy. *Transactions of the New York Academy of Sciences*, *39*, 147–157. doi: 10.1111/j.2164-0947.1980.tb02775.x

Su, Y., & Cambanis, S. (1993). Sampling designs for estimation of a random process. *Stochastic Processes and their Applications*, *46*, 47–89.

Takougang, I., Meremikwu, M., Wanji, S., Yenshu, E., Aripko, B., Lamlenn, S., ... Remme, J. (2002). Rapid assessment method for prevalence and intensity of loa loa infection. *Bulletin of the World Health Organ*, *80(11)*, 852-8.

Tangpukdee, N., Duangdee, C., Wilairatana, P., & Krudsood, S. (2009). Malaria diagnosis: a brief review. *The Korean journal of parasitology*, *47*, 93.

Taylor, B., Davies, T., Rowlingson, B., & Diggle, P. (2013). LGCP: Inference with spatial and spatio-temporal log-Gaussian Cox processes in R. *Journal of Statistical Software*, *52*, 1–40. doi: 10.18637/jss.v052.i04

Tene Fossog, B., Ayala, D., Acevedo, P., Kengne, P., Ngomo Abeso Mebuy, I., Makanga, B., ... Costantini, C. (2015). Habitat segregation and ecological character displacement in cryptic African malaria mosquitoes. *Evolutionary Applications*, *8*, 326–345. doi: 10.1111/eva.12242

Tobler, W. (1970). A computer movie simulating urban growth in the Detroit region. *Economic Geography*, *46*, 234–240.

Varin, C., Reid, N., & Firth, D. (2011). An overview of composite likelihood methods. *Statistica Sinica*, *21*, 5–42.

Vecchia, A. V. (1988). Estimation and model identification for continuous spatial processes. *Journal of the Royal Statistical Society, Series B*, *50*, 297–312.

Wanji, S., Akotshi, D., Kankou, J., Nigo, M., Tepage, F., Ukety, T., ... Remme, J. (2012). The validation of the rapid assessment procedures for loiasis (RAPLOA) in the Democratic Republic of Congo: health policy implications. *Parasites & Vectors*, *5*, 25. doi: 10.1186/1756-3305-5-25

Warrick, A., & Myers, D. (1987). Optimization of sampling locations for variogram calculations. *Water Resources Research, 23*, 496–500.

Watson, G. (1971). Trend -surface analysis. *Mathematical Geology, 3*, 215–226.

Watson, G. (1972). Trend surface analysis and spatial correlation. *Geological Society of America Special Paper, 146*, 39–46.

Weisberg, S. (2013). *Applied Linear Regression* (fourth ed.). Wiley.

Whittle, P. (1953). The analysis of multiple stationary time series. *Journal of the Royal Statistical Society, Series B, 15*, 125-139.

Whittle, P. (1963). Stochastic processes in several dimensions. *Bulletin of the International Statistical Institute, 40*, 974–994.

Xie, Y. (2013). animation: An R package for creating animations and demonstrating statistical methods. *Journal of Statistical Software, 53*, 1–27.

Yfantis, E., Flatman, G., & Behar, J. (1987). Efficiency of kriging estimation for square, triangular and hexagonal grids. *Mathematical Geology, 19*, 183–205.

Zhu, Z., & Stein, M. (2006). Spatial sampling design for prediction with estimated parameters. *Journal of Agricultural, Biological, and Environmental Statistics, 11*, 24–44.

Zimmerman, D. (2006). Optimal network design for spatial prediction, covariance parameter estimation, and empirical prediction. *Environmetrics, 17*, 635–652.

Zouré, G. M., Honorat, Noma, M., Tekle, H., Afework, Amazigo, U. V., Diggle, P. J., Giorgi, E., & Remme, J. H. F. (2014). The geographic distribution of onchocerciasis in the 20 participating countries of the african programme for onchocerciasis control: (2) pre-control endemicity levels and estimated number infected. *Parasites & Vectors, 7*. doi: 10.1186/1756-3305-7-326

Index